Viktor Dubec

Optical Testing of Semiconductor Devices under High Energy Pulses

Viktor Dubec

Optical Testing of Semiconductor Devices under High Energy Pulses

Advanced Optical Interferometric Methods for Nanosecond Mapping of Semiconductor Devices under High Energy Pulses

Südwestdeutscher Verlag für Hochschulschriften

Impressum/Imprint (nur für Deutschland/ only for Germany)
Bibliografische Information der Deutschen Nationalbibliothek: Die Deutsche Nationalbibliothek verzeichnet diese Publikation in der Deutschen Nationalbibliografie; detaillierte bibliografische Daten sind im Internet über http://dnb.d-nb.de abrufbar.
Alle in diesem Buch genannten Marken und Produktnamen unterliegen warenzeichen-, marken- oder patentrechtlichem Schutz bzw. sind Warenzeichen oder eingetragene Warenzeichen der jeweiligen Inhaber. Die Wiedergabe von Marken, Produktnamen, Gebrauchsnamen, Handelsnamen, Warenbezeichnungen u.s.w. in diesem Werk berechtigt auch ohne besondere Kennzeichnung nicht zu der Annahme, dass solche Namen im Sinne der Warenzeichen- und Markenschutzgesetzgebung als frei zu betrachten wären und daher von jedermann benutzt werden dürften.

Verlag: Südwestdeutscher Verlag für Hochschulschriften Aktiengesellschaft & Co. KG
Dudweiler Landstr. 99, 66123 Saarbrücken, Deutschland
Telefon +49 681 37 20 271-1, Telefax +49 681 37 20 271-0, Email: info@svh-verlag.de
Zugl.: Wien, TU, Diss., 2005

Herstellung in Deutschland:
Schaltungsdienst Lange o.H.G., Berlin
Books on Demand GmbH, Norderstedt
Reha GmbH, Saarbrücken
Amazon Distribution GmbH, Leipzig
ISBN: 978-3-8381-0404-1

Imprint (only for USA, GB)
Bibliographic information published by the Deutsche Nationalbibliothek: The Deutsche Nationalbibliothek lists this publication in the Deutsche Nationalbibliografie; detailed bibliographic data are available in the Internet at http://dnb.d-nb.de.
Any brand names and product names mentioned in this book are subject to trademark, brand or patent protection and are trademarks or registered trademarks of their respective holders. The use of brand names, product names, common names, trade names, product descriptions etc. even without a particular marking in this works is in no way to be construed to mean that such names may be regarded as unrestricted in respect of trademark and brand protection legislation and could thus be used by anyone.

Publisher:
Südwestdeutscher Verlag für Hochschulschriften Aktiengesellschaft & Co. KG
Dudweiler Landstr. 99, 66123 Saarbrücken, Germany
Phone +49 681 37 20 271-1, Fax +49 681 37 20 271-0, Email: info@svh-verlag.de

Copyright © 2009 by the author and Südwestdeutscher Verlag für Hochschulschriften Aktiengesellschaft & Co. KG and licensors
All rights reserved. Saarbrücken 2009

Printed in the U.S.A.
Printed in the U.K. by (see last page)
ISBN: 978-3-8381-0404-1

Kurzfassung

Die Zuverlässigkeit von Halbleiterbauelementen ist der erste Schritt zu einem sicheren Betrieb von elektronischen Schaltungen. Zur Optimierung von Bauelementen und zur Überprüfung von Bauteilsimulationsmodellen ist das Wissen der Wärme- und der freien Ladungsträgerverteilung in Bauelementen essentiell. Die Temperaturüberwachung ist für Bauelemente, welche mit hohen Stromdichten arbeiten, sehr wichtig, da Selbsterwärmung bei diesen die Hauptfehlerursache ist.

Zur Untersuchung von transienten Temperaturverteilungen oder Änderungen der Verteilungen der freien Ladungsträger in Bauelementen sind zerstörungsfreie optische Methoden, welche auf der Beobachtung des Brechungsindex, Absorption oder Lichtemission basieren, bekannt. Jedoch leiden diese Methoden entweder an geringer Ortsauflösung oder geringer Zeitauflösung. Die überlagernde abtastende Methode wurde kürzlich für Untersuchungen von dynamischen Stromverteilungen während kurzer elektrischer Pulse verwendet, jedoch hat sie keine ausreichende Zeitauflösung für Untersuchungen im CDM (charged device model) Zeitbereich. Ferner benötigt sie wiederholte Belastungen der untersuchten Bauelemente. Dadurch kann das Bauelement zerstört werden oder einige der Bauelementeigenschaften - wie z.B. Puls zu Puls Instabilität - kann unerkannt bleiben. Deshalb wurden zwei Untersuchungsmethoden aufgebaut, welche auf transienter interferometrischer Abbildung (transient interferometric mapping - TIM) beruhen. Ersteres basiert auf einem zu mehreren Zeitpunkten zweidimensionalen Einzelschussverfahren und Zweiteres beruht auf einer Zweistrahlmethode mit einer Zeitauflösung im Subnanosekunden Bereich.

Die zweidimensionale TIM-Methode liefert Informationen über die Stromverteilung in Bauelementen zu zwei Zeitpunkten während eines einzigen elektrischen Belastungspulses. Dieser Arbeit umfasst die Entwicklung und Optimierung des optischen zweidimensionalen TIM Messaufbaues und die Messdatenanalyse. Der Messaufbau ermöglicht die Untersuchung von einzelnen Halbleiterchips und ganzen Halbleiterkristallscheiben. Als beste Methode zur Datenanalyse hat sich eine auf der Fourier-Transformation basierende Lösung herausgestellt. Es wurde gezeigt, dass das Bauelementlayout die Phasenverteilung beeinflusst und Verschiebungen in der Phase hervorruft. Deshalb wurde die Filterung des Spektrums im Detail untersucht, um die

Kurzfassung

Welligkeit und das Rauschen zu minimieren. Mit Hilfe von Simulationen wurde ein adaptiver Filter für das Spektrum entwickelt. Mehrere Methoden zur Phasenrekonstruierung wurden überprüft und an deren Analyse wurde eine Phasenvoraufbereitung entwickelt. Diese beschleunigt nicht nur die Phasenrekonstruierung, sondern eliminiert bzw. isoliert Störobjekte in der Phasenverteilung und reduziert den Bedarf nach manuellen Phasenkorrekturen auf kleine Bereiche. Die zweidimensionale TIM Methode ermöglicht ebenso die Gewinnung der momentanen zweidimensionalen Verlustleistungsdichte. Ferner wurde der Einfluss der Bauelementtopologie, der Kamera und des Lasersystems auf das Interferogramm im Detail analysiert, um die Messgenauigkeit abzuschätzen. Der Messaufbau wurde erfolgreich zur Untersuchung von bewegten Stromfilamenten in gekoppelten npn/pnp ESD Schutzstrukturen, von instabilen Stromverteilungen und von zerstörenden Phänomenen in selbstgeschützten vertikalen DMOS Transistoren angewandt. Es wurde weiters gezeigt, dass der Messaufbau auch Abbildung der Temperaturverteilung nach der thermisch induzierten Änderung der Lichtabsorption in Halbleitern mit einer zeitlichen Auflösung im Nanosekundenbereich verwendet werden kann.

Für Untersuchungen im CDM Zeitbereich wurde ein weiterer Messaufbau, welcher auf einem Michelson Interferometer basiert, mit einer Zeitauflösung im Subnanosekundenbereich entwickelt. Unter Verwendung von zwei fokussierten Laserstrahlen konnte die absolute Phasenverschiebung an zwei unterschiedlichen Positionen im Bauelement gemessen werden. Der Messaufbau wurde dahingehend optimiert, um möglichst geringe elektromagnetische Störungen von den Flanken des Hochleistungsbelastungspulses zu erhalten. Weiters wurde er zur Untersuchung von Auslöseverzögerungen und inhomogenen Stromverteilungen in ESD Schutzstrukturen angewandt.

Abstract

Reliability of semiconductor devices is the first step for safe operation of electronic circuits. For optimisation of devices and for verification of device simulation models the knowledge of heat dissipation and of free carrier concentration in the device is essential. Temperature monitoring is especially important for devices operating at high current densities, where self-heating is a main failure cause.

For investigation of transient temperature or free carrier changes within the devices, non-destructive optical methods based on monitoring of the refractive index, absorption or light emission have previously been developed. However, these methods suffer either from small spatial or time resolution. The heterodyne scanning technique has previously been introduced for investigation of the current dynamics during short electrical pulses; however, it has not sufficient time resolution for investigation in CDM (charged device model) time domain and it requires repeatable stressing of the device. As a result, the device can either be destroyed or some of the device features like trigger pulse-to-pulse instability may be hidden. Therefore two testing techniques based on transient interferometric mapping (TIM) have been developed within this thesis: the two-dimensional (2D) multiple-time-instant single-shot technique and the two-beam technique with sub-nanosecond time resolution.

The 2D TIM technique provides information about the current flow distribution in the device at two time instants during a single electrical stress pulse. One goal of this work was to develop and optimise the 2D TIM setup optical layout and the data analysis method. The setup enables testing of single chips and also on wafer level. The best method for the data analysis was found to be the Fourier transform based technique. It was shown that the device layout influences the final phase distribution and induces phase undulations. Therefore the spectrum filtering was studied in detail in order to limit the undulations and noise and with the help of simulation an adaptive spectrum filter was proposed. Various phase unwrapping methods were examined and based on this analysis a phase pre-processing was proposed, which not only speeds up the phase unwrapping process but also eliminates and isolates the phase artifacts and reduces the need for unwrapping to a local area. The 2D TIM technique enables also extraction of the instantaneous 2D power dissipation density. Furthermore, to estimate the measurement

Abstract

accuracy, the effect of the device topology, camera and laser properties on the interferogram was analysed in detail. The setup was successfully applied to study the moving current filaments in coupled npn/pnp ESD protection devices and to study current flow instability and destructive phenomena in self-protecting vertical DMOS transistors. It was also demonstrated that the setup can be used for the thermal imaging using the temperature-induced changes of light absorption in the semiconductor bulk with nanosecond time resolution.

For investigation in CDM time domain a setup based on the Michelson interferometer with sub-nanosecond time resolution was developed. Using two focused laser beams the absolute phase shift at two different positions on the device can be measured. The setup was optimised to avoid the electromagnetic pick-ups arising from the rising/falling edge of the high power stress pulses. The setup was applied to study short trigger delays and inhomogeneous current flow in ESD protection devices.

Content

1 INTRODUCTION 1

1.1 MOTIVATION 1
1.2 OUTLINE OF THE THESIS 2
1.3 ELECTROSTATIC DISCHARGE 3
1.3.1 ESD models 3
1.4 OPTICAL CHARACTERISATION OF DEVICES 4
1.4.1 Overview 4
1.4.2 Optical beam probing 5
1.4.2.1 Complex refractive index change 6
1.4.3 Backside laser interferometry (BLI) 8
1.4.4 Concepts of backside laser interferometry 9
1.4.5 Holographic interferometry 10
1.4.6 Errors in the interferogram 12
1.4.7 Sign ambiguity, 2π-uncertainty 13
1.4.8 Interferogram evaluation methods – overview 14
1.4.8.1 Fringe skeletonizing 14
1.4.8.2 Temporal heterodyning 15
1.4.8.3 Phase sampling 15
1.4.8.4 Fourier transform evaluation 16
1.4.9 Comparison of phase extraction methods 20
1.5 PHASE UNWRAPPING 21
1.5.1 Straightforward algorithm 23
1.5.2 Spiral scanning algorithm 23
1.5.3 Pixel queue algorithm 24
1.5.4 Minimum spanning tree algorithm 25
1.5.5 Cellular automata 26
1.5.6 Other algorithms and approaches 26
1.5.7 Comparison of phase unwrapping methods 27

    1.5.8 Error sources in the unwrapping process 28
  1.6   2D POWER DISSIPATION DENSITY 29

2 2D SETUP 31

  2.1   SETUP DESCRIPTION 31
    2.1.1 Imaging at single time instant 31
    2.1.2 Imaging at two time instants 33
        2.1.2.1   Delay line variant 33
        2.1.2.2   Two lasers variant 34
    2.1.3 Probe station 36
  2.2   DETECTION SCHEME 38
    2.2.1 Measurement framework 38
    2.2.2 Triggering and data acquisition 39
  2.3   SETUP SPECIFICATIONS 41
    2.3.1 Field of view 41
    2.3.2 Spatial resolution 42
    2.3.3 Time resolution 43
    2.3.4 Laser specifications 44
    2.3.5 Camera specifications 45
    2.3.6 Electrical stressing 47
  2.4   EFFECT OF THE DUT ON THE MEASUREMENT 48
    2.4.1 Sample backside surface roughness (polishing) 48
    2.4.2 Reflectivity of the sample 50
    2.4.3 Lateral geometry of the device and fringe discontinuity 52
    2.4.4 Cellular device structure and the phase extraction accuracy 53
  2.5   PHASE EXTRACTION 55
    2.5.1 Phase extraction sequence 56
    2.5.2 Spectrum filtering 59
        2.5.2.1   Noise filtering 60
        2.5.2.2   Spectrum overlapping 63
        2.5.2.3   Optimal filter design 69
    2.5.3 Phase unwrapping 72
        2.5.3.1   Phase pre-processing 75

2.5.4 Error sources of the phase measurement 76
    2.5.4.1 Effect of finite pulse duration 76
    2.5.4.2 Multiple reflections in the DUT 77
    2.5.4.3 Effect of inhomogeneous laser intensity distribution 83
    2.5.4.4 Pulse to pulse instabilities of the laser beam 84
2.6 CONCLUSION – OPTIMAL MEASUREMENT PARAMETERS 86
2.7 2D POWER DISSIPATION DENSITY 87
    2.7.1 Optimal delay δt *92*
2.8 EXAMPLES 92
    2.8.1 Example 1 – Phase calibration *92*
    2.8.2 Example 2 – Moving current filament *93*
    2.8.3 Example 3 – Unrepeatable device triggering in power DMOS *95*
    2.8.4 Example 4 – Destructive phenomena measurement *97*
2.9 THERMAL IMAGING USING ABSORPTION MEASUREMENTS 98
    2.9.1 Example – Spreading current filament *99*
2.10 SETUP FURTHER DEVELOPMENT 100

3 DUAL-BEAM INTERFEROMETER 103

3.1 INTRODUCTION 103
3.2 SETUP DESCRIPTION 103
    3.2.1 Optical layout *103*
    3.2.2 Setup parameters *105*
    3.2.3 Electrical device testing *106*
        3.2.3.1 TLP technique 106
        3.2.3.2 vf-TLP technique 107
3.3 PHASE MEASUREMENT AND CALCULATION 108
    3.3.1 Signal timing *109*
3.4 ERROR SOURCES 110
    3.4.1 Vibrations *111*
    3.4.2 Optical feedback *111*
    3.4.3 Electromagnetic pick-ups *112*
3.5 EXAMPLES 114
    3.5.1 Example 1 – Phase calibration *114*

    3.5.2   Example 2 – Separation of thermal and free carrier contribution 115

    3.5.3   Example 3 – Measurement of small trigger delay 116

    3.5.4   Example 4 – Measurement of unrepeatable phenomena 117

  3.6   SUMMARY 118

4 SUMMARY 120

APPENDIX 122

BIBLIOGRAPHY 127

LIST OF PUBLICATIONS AND CONFERENCE CONTRIBUTIONS 137

LIST OF ACRONYMS 140

LIST OF FREQUENTLY USED SYMBOLS 142

ACKNOWLEDGEMENT 144

1 Introduction

1.1 Motivation

Nowadays society becomes more and more dependent on the micro-electronic circuits that are implemented in the daily used devices. These devices can be consumable electronics as well as part of automotive systems, security electronics, military devices, medicine instruments or communication sector. Because the failure of these circuits can lead up to the life-jeopardy situation, their reliability becomes more and more important.

The most probable reason of the device failure is the self-heating. The high current in the device, which can be caused e.g. by an electrostatic discharge (ESD), can lead to the device local overheating and destruction. For the device designers it is therefore important to get the knowledge about the heat dissipation in the device. This knowledge can be obtained from the device simulation or from the device testing.

In the simulation process the device internal behaviour is predicted. The simulation result is compared with the measured IV characteristics obtained by the measurement. However, the comparison of voltage waveforms does not verify the internal behaviour. Furthermore, the uncertain physical models of the semiconductors at high temperatures, the three-dimensional effects and the doping profile uncertainty limit the simulation accuracy. Models for physical effects like an avalanche multiplication rate in the high temperature regime have still to be developed.

The methods for the experimental testing of the device internal behaviour can be split into destructive and non-destructive group. The destructive methods require a large number of tested devices and it is an irreversible process, which is very uncomfortable. The non-destructive methods can be based e.g. on optical beam testing. These non-destructive techniques provide information about the thermal distribution and current flow distribution in the device, which is not measurable by any other methods. Therefore their development is important for the industry. These methods either measure the signal on the device surface, but these miss the temporal or spatial resolution, or they measure the signal directly inside the device, accessing the device from the backside. The backside access is necessary due to the multilevel metal composition on the topside in the nowadays technology.

1 Introduction

One of the well-known backside optical testing methods is the heterodyne technique. The advantages of this technique are the high sensitivity to the small temperature changes and the good spatial and temporal resolution. However this technique is not convenient for the measurement of the destructive phenomena and non-repetitive device triggering. This is possible to measure by a new two-dimensional (2D) optical testing method based on the interferometric mapping of the refractive index changes and absorption changes, which has been developed and which is in the focus of this work. The method speeds up the testing process and enables a direct insight into the device behaviour, which is not possible by any other method. It provides information about the current flow distribution in the device at any time instant for a single electrical stress pulse.

Mostly in the automotive industry the understanding of the device behaviour at nanosecond time scale is important for the device optimisation. Within this work a new setup based on the Michelson interferometer is presented. This setup has a sub-nanosecond time resolution and thus it is a unique tool for mapping of current flow dynamics within the device under the fast events like ESD issuing from charged instruments.

1.2 Outline of the thesis

The purpose of this work was to develop and characterise two optical setups for measurement of the refractive index changes and absorption changes in the semiconductor devices under the short electrical pulse in the industrial-like conditions. The thesis is split into two parts regarding of two optical setups.

In the first part a backside 2D transient interferometric mapping (TIM) technique is presented. The technique is based on the principles of holographic interferometry in reflection mode and uses the thermo-optical effect. The work emanates from a preliminary concept of the setup. The main effort was concentrated on the improvement of the setup hardware, sensitivity of the measurement, to evaluate the effect of device on the measurement, to interpret and examine the errors appearing in the result, to analyse various methods for the result evaluation in order to improve the measurement sensitivity and automation of the measurement process for purpose of implementation in the industry.

1 Introduction

In the second part of this thesis a setup based on a Michelson interferometer with a sub-ns time resolution is presented. The setup allows measurement of the heat dynamics and free carrier density changes in a single point or two points in the device simultaneously. The examples demonstrate the main features and the gain from such measurements, which can not be provided by any other tool.

1.3 Electrostatic discharge

The ubiquitous electrostatic discharge (ESD) is one of the most probable reasons of the fault of the circuits. Therefore the circuits have to be well protected against the ESD. The ESD can occur during the production and during the usage of the device. It is introduced by a person or by instrument in the vicinity. Touching the device or even coming near to the device can create an ESD event. The temperature in the device during the ESD event can rise up to the critical levels and cause an irreversible harm of the device. The nowadays device down scaling is accompanied by increased doping levels and leads to the increased sensitivity to the ESD phenomena. To avoid the device destruction, the ESD protection structures are implemented into the circuits. The lifetime of the device depends then on the reliability and ruggedness of this protection structure.

The ESD protection structure has to protect every input-output pin of the circuit. It provides a discharge path for the high currents and limits the high voltage at the contact pads. Because the protection structure has to serve reliably but may not influence the functionality of the whole circuit, to design the protection structure it is necessary to understand its internal behaviour under the ESD stress and its interaction with the circuit.

1.3.1 ESD models

Two ESD events differ in duration, power and waveform depending on the environment conditions. Therefore the leading organisations in the industry, which investigate the ESD event, have created standard models for ESD testing [ESD-R, MIL-STD, EIA/JEDEC, Ame95].

1 Introduction

There are two main models of the ESD event. The most important and widest used is the human body model (HBM). It simulates the ESD event produced by a charged human body touching the electronic circuit. The typical risetime of this discharge is 2-10 ns, duration around 150 ns, amplitude of up to few amperes, see Fig. 1.1a. It is mainly responsible for the thermal destruction of the junction and contacts due to self-heating effect [Kel96].

With the increasing automation in the industry the charged device model (CDM) becomes more and more important [Ame92]. The machine that handles the chip can be charged due to movement and friction. When the machine touches the chip, a short but high power discharge stresses the chip, mostly resulting in electrical breakdown of the gate oxide [Rei95, Mal88]. The typical risetime of CDM pulse is below 1 ns, duration around 3 ns and amplitude of several amperes, see Fig. 1.1b.

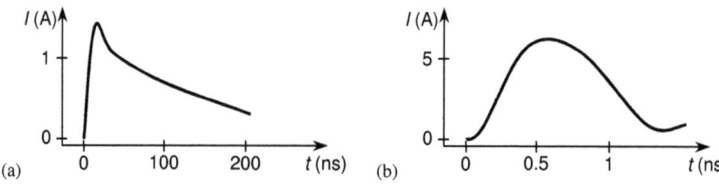

Fig. 1.1. The ESD models: (a) the HBM pulse and (b) the CDM pulse, after [Ame95].

1.4 Optical characterisation of devices

1.4.1 Overview

The optical methods for the non-destructive device inspection are based on various principles. Some of them are commercially used in the industry.

Optical pyrometry (infrared thermography) [Her95a] is based on the detection of the black body radiation with spatial resolution of several micrometers, sensitivity of 1 K and temporal resolution of 10 µs [Kol96, Bre97].

High sensitivity to the temperature up to 0.1 K is achieved in liquid crystal thermography by application of a liquid crystal on the device topside [Fer68, Cse95, Ver94]. The increased temperature in the layer leads to the change in the polarisation of the incident light, when this light passes through the crystal layer. The disadvantage of these methods is the ms time resolution. Space resolution is a few μm.

Similar method is a fluorescent microthermal imaging, where a fluorescent material layer is applied on the device top surface [Gla96, Bar95, Her98, Kol96]. The layer illuminated by an UV light generates a fluorescence spectrum in the visible range, which is temperature dependent. The spatial resolution is around 0.3 μm and the temperature sensitivity 10 mK.

The light emission microscopy is based on the detection of the radiation, which is irradiated during the electron-hole recombination or the field-acceleration of a charged particle [Deb93, Kol92, Lun91a, Rus98]. From the spectrum analysis different mechanisms can be resolved, since they are correlated to different spectral characteristics. The usage of this method for the silicon devices is limited by absorption of the radiation in the substrate.

Based on time resolved light emission microscopy a method called picosecond imaging circuit analysis (PICA) has been introduced, which records the time and position of individual photons [Rem03]. The time resolution of 100 ps is achieved with a gated intensified camera. The main drawback is the long acquisition time (several hours) due to the detector poor quantum efficiency.

1.4.2 Optical beam probing

Large group of device testing methods is based on optical beam probing [Ble92, Lun91b, Pan98]. These methods monitor the changes of the refractive index or the absorption coefficient (simply complex refractive index) of the material due to the free carrier concentration changes (plasma-optical effect) [Sor87, Stu92] or temperature changes (thermo-optical effect) [McC94, Ber90, Ice76, Her94].

A method based on probing with a laser light is Fabry-Perot interferometry [Pog98b, Don90]. The device is illuminated by a coherent light and the multiple reflections inside the substrate result to an interference pattern. Due to the thermo-optical and thermo-mechanical effect the optical path changes with the rising temperature and the temperature can be extracted.

Another optical beam probing method is based on the Schlieren imaging [Sch01]. A measurement of the heat-induced refractive index gradient changes is used here for the hot spot detection in devices [Nie02].

1 Introduction

1.4.2.1 Complex refractive index change

The variation of the complex refractive index with the temperature is a consequence of variation of the bandgap energy with the temperature and phonon assisted absorption processes [Kos88]. The variation of the complex refractive index with the free carrier concentration results from the Maxwell equations and Drude theory of free electrons [Bor80, Sor87].

The refractive index change Δn can be divided into a thermal contribution Δn_{th}, free electron contribution $\Delta n_{fc,e}$ and free hole contribution $\Delta n_{fc,h}$:

$$\Delta n(x,y,z,t) = \Delta n_{th}(x,y,z,t) + \Delta n_{fc,e}(x,y,z,t) + \Delta n_{fc,h}(x,y,z,t) \tag{1.1}$$

where:

$$\Delta n_{th}(x,y,z,t) = n_{th}(T(x,y,z,t)) - n_{th}(T_0)$$

$$\Delta n_{fc,e}(x,y,z,t) = k_e [c_e^{\alpha'}(x,y,z,t) - c_e^{\alpha'}(x,y,z,t_0)] \tag{1.2}$$

$$\Delta n_{fc,h}(x,y,z,t) = k_h [c_h^{\beta'}(x,y,z,t) - c_h^{\beta'}(x,y,z,t_0)]$$

Here n_{th} is the refractive index at temperature T, T_0 is the ambient temperature, c_e and c_h are the electron and hole concentrations at spatial coordinates x, y, z and time t, t_0 is the time corresponding to the steady-state. The remaining parameters α', β', k_e, k_h are coefficients obtained from the fitting of experimental data. More details about the $\Delta n_{fc,e}$, $\Delta n_{fc,h}$ and coefficients α', β', k_e, k_h are in [Sor87].

The refractive index Δn_{th} is temperature dependent. The dependence of Δn_{th} for silicon according to [McC94] is plotted in Fig. 1.2. The dependence can be considered to be linear in first approximation. The relation for Δn_{th} is then:

$$\Delta n_{th} = \frac{dn}{dT} * (T(x,y,z,t) - T_0), \tag{1.3}$$

where dn/dT is the temperature coefficient of the refractive index (for Si at 300 K $dn/dT = 0.00019$ K^{-1}).

1 Introduction

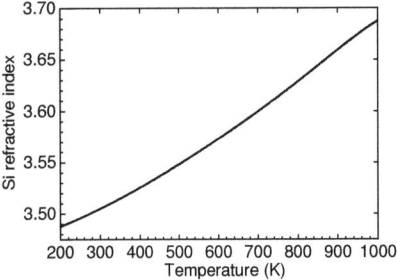

Fig. 1.2. Temperature dependence of silicon refractive index according to [McC94].

The absorption coefficient α can be in first approximation expressed as a sum of thermal and free carrier contribution α_{th} and α_{fc}:

$$\alpha(T, c_e, c_h) = \alpha_{th}(T) + \alpha_{fc}(c_e, c_h) \qquad (1.4)$$

where:

$$\alpha_{th}(T) = \alpha_{th,bg}(T) + \alpha_{th,fc}(T)$$
$$\alpha_{fc}(c_e, c_h) = a_e c_e^{\delta}(x, y, z, t) + a_h c_h^{\gamma}(x, y, z, t) \qquad (1.5)$$

Here $\alpha_{th,bg}$ is a band-to-band absorption coefficient (see [Tim93] for details) and $\alpha_{th,fc}$ is a part related to a temperature increase in the free carrier absorption due to the increase in the intrinsic concentration [Rog96]. Parameters a_e, a_h, δ, γ are obtained from the fitting of experimental data in [Sor87]. The dependence of α_{th} for silicon according to [Rog96, Tim93] is plotted in Fig. 1.3.

The change of the absorption coefficient α causes a change in the optical intensity from I_0 to I. If the light passes an area of length L, where the absorption coefficient α vary, the intensity of the transmitted light is:

$$I(x, y) = I_0 \exp\left(-\int_0^L \alpha(x, y, z) dz\right). \qquad (1.6)$$

1 Introduction

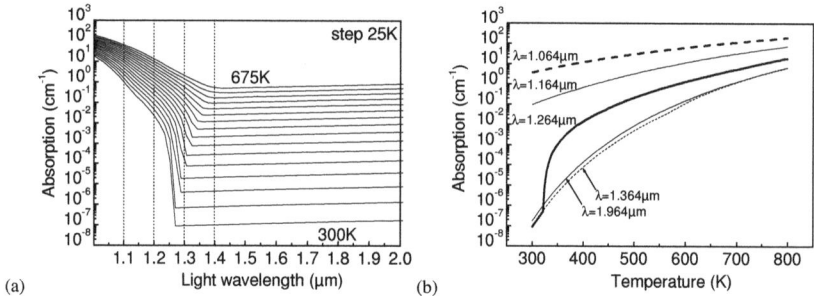

Fig. 1.3. (a) Dependence of silicon absorption coefficient on the light wavelength for different temperatures. (b) Temperature dependence of the silicon absorption coefficient for different wavelengths calculated after [Rog96, Tim93]

1.4.3 Backside laser interferometry (BLI)

A group of optical probing methods is based on backside laser interferometry (BLI). The device is illuminated from the substrate side with a probe laser beam, for which the substrate is transparent. The phase of the probe beam is thus modified by the transient refractive index of the substrate. The main advantage of the phase measurement methods against other optical probing methods is the high dynamic range.

The probe beam can be either focused to a point [Hei86b] or to stay wide to illuminate the whole DUT [Kre96]. The probe beam passes through the substrate and is reflected back from the device topside, see Fig. 1.4. Thus the probe beam carries the information about the refractive index change Δn along its path. This information is extracted using the interference with a reference beam. The transient phase shift $\Delta\varphi$ of the probe beam due to Δn is expressed as:

$$\Delta\varphi(x,y,t) = 2\frac{2\pi}{\lambda}\int_0^L \Delta n(x,y,z,t)dz \qquad (1.7)$$

8

1 Introduction

where λ is the laser beam wavelength. The factor 2 originates from the fact, that the beam passes the substrate twice. According to Eq. 1.1, the phase shift can be expressed as a sum of thermo-optical φ_{th} and plasma-optical φ_{fc} component:

$$\Delta\varphi = \varphi_{th} + \varphi_{fc} \qquad (1.8)$$

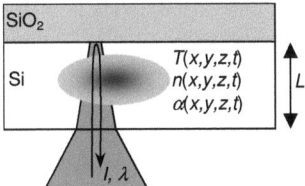

Fig. 1.4. The principle of the backside interferometric probing. The laser beam passes the substrate and is reflected back from the device topside.

The validity of Eq. 1.7 is limited by two restrictions: the Δn has to be finite and the multiple reflections inside the device have to be avoided [Sel97, Pog02c, Pog97]. The first restriction is fulfilled if the Δn originates from thermo-optical effect, where the temperature does not change abruptly on a distance of size bellow the wavelength. The second restriction is fulfilled if the beam is focused with an objective of high numerical aperture or a laser with short coherent length is used. In opposite case the optical matrix or transmission line formalism has to be applied.

1.4.4 Concepts of backside laser interferometry

One of the methods based on BLI uses a linearly polarised beam, which is split into reference and probe beam of orthogonal polarisation [Bla87, Hei86a, Kra92]. The reflected probe beam with modified phase is composed with a reference beam into one elliptically polarised beam, which polarisation change is detected. This method is sensitive to small temperature differences.

Another concept is based on a Michelson interferometer, where the temperature induced phase difference directly reflects into the intensity change of the interference signal. This

1 Introduction

interference signal is composed from one reference and one probe beam. The probe beam is focused in one point, where the optical signal is measured. This is commercially used e.g. in IDS 2500 probe system from Schlumberger for measuring of the transistors in flip-chip packages. The advantage of this concept is the high speed.

In the heterodyne interferometer an acoustic-optic modulator is used to split an infrared (IR) laser beam into a probe $I_p(t)$ and reference beam $I_r(t)$ with a little frequency difference [Gol93, Fur99]. The interference of such two beams results into a heterodyne beating signal measured by an optical detector. The advantage of this setup is the high phase sensitivity, simple automation and that the detected signal is not disturbed by beam amplitude modulation.

The main disadvantage of the mentioned scanning setups is that the device has to be repetitively stressed. This excludes possibility of measurement near to the damage threshold, measurement of destructive events and measurement of devices exhibiting non-repetitive behaviour from pulse to pulse. In addition, the scanning process is a time consuming procedure. All this disadvantages are excluded in method based on a holographic interferometry [Kre96], which is introduced in the next chapter.

1.4.5 Holographic interferometry

In holographic interferometry a broad probe beam illuminates the whole device area and therefore contains the information about the phase profile of the whole object. This is then recorded via a 2D recording medium (e.g. CCD camera [Sch95, Yam96]) together with a reference beam, resulting thus into a holographic interference pattern, called also holographic interferogram [Kre96].

In the holographic interferometry, two plane waves E_1 and E_2 (reference and probe wave) of the same frequency ω but of different wave vectors $\vec{k}_1$ and $\vec{k}_2$ and different phase φ_1 and φ_2 interfere:

$$E_1(\vec{r},t) = E_{01}(\vec{r})e^{i(\vec{k}_1 \cdot \vec{r} - \omega t + \varphi_1(\vec{r}))}$$
$$E_2(\vec{r},t) = E_{02}(\vec{r})e^{i(\vec{k}_2 \cdot \vec{r} - \omega t + \varphi_2(\vec{r}))} \qquad (1.9)$$

The intensity I in any point in the plane of detection is as follows:

1 Introduction

$$I(\vec{r}) = |E_1(\vec{r},t) + E_2(\vec{r},t)|^2 = E_{01}^2(\vec{r}) + E_{02}^2(\vec{r}) + 2E_{01}(\vec{r})E_{02}(\vec{r})\cos\left[\left(\vec{k}_1 - \vec{k}_2\right)\vec{r} + \varphi_1(\vec{r}) - \varphi_2(\vec{r})\right]$$
(1.10)

which can be written as:

$$I(x,y) \equiv A(x,y) + B(x,y)\cos[\phi(x,y)]$$
(1.11)

This time-independent pattern is called interferogram (or interference pattern) and the fringes interference fringes, see example in Fig. 1.5. The first term A represents the background intensity, the second term B the fringe amplitude. The distance of the fringes for two plane waves is $2\pi/|\vec{k}_1 - \vec{k}_2|$ in the direction of vector $\vec{k}_1 - \vec{k}_2$.

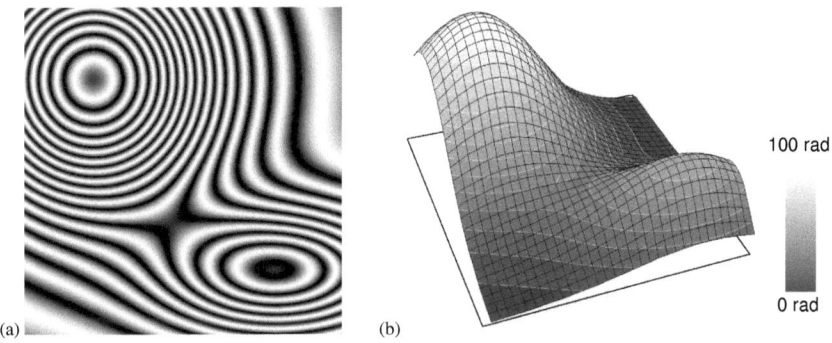

Fig. 1.5. (a) Simulated interferogram $I(x, y)$ and (b) corresponding phase $\phi(x, y)$.

In real conditions, where the light source is not an ideal source of coherent light, the interference term has to be multiplied by the normalised spatio-temporal coherence function (degree of coherence) $\gamma(s,\tau)$ [Bor80, Sal91]:

$$I(x,y) = A(x,y) + \gamma(s,\tau)B(x,y)\cos[\phi(x,y)]$$
(1.12)

where s is the spatial shift of the same point of the two interfering waves in plane of observation and τ is the temporal shift of the waves.

1.4.6 Errors in the interferogram

The interferograms picked up by a 2D sensor suffer from a number of distortions, which degrade the interference patters and thus complicate the phase extraction [Kre96, Vro91, Kem03]. Some of the distortions can not be avoided during recording and depending on the nature of the errors the right choice of the interferogram evaluation method is important.

The Eq. 1.11 describes a general interferogram. Here the terms A and B are not constant over the interferogram area due to (i) the object reflectivity variation (e.g. metals reflect more light than silicon), (ii) non-uniform laser beam intensity profile (e.g. Gaussian profile) and (iii) non-uniform sensor sensitivity (e.g. vidicon camera is more sensitive in the middle that at the edges of the tube). Additionally B is influenced (i) by the spatio-temporal coherence of the light (see Eq. 1.12) and (ii) by the speckles [Ras94, Jon89, Kad97]. The term A includes also (i) the electronic noise of the recording medium (thermal noise, shot noise, generation-recombination noise, 1/f noise, photon noise), (ii) the diffraction patterns arising from the dust particles in the optics and apertures and (iii) the stray reflections from the optics.

Moreover, the environmental distortions degrade the interferograms. These are mechanical vibrations in the setup, acoustic noise, which is also a source of the vibrations, and the air turbulence, which may cause refractive index change of the air leading thus to different optical path between the two interfering beams. These effects play role for long exposure times or if two sequentially recorded interferograms have to be compared.

The topology of the illuminated object can degrade the interferogram by introducing discontinuities to the fringes and breaking or splitting of the fringes. In addition, steep edges of the objects can introduce closely spaced fringes, which complicates the evaluation. Extraneous fringes coming from the multiple reflections disturb the fringe pattern too. Fault detection in the fringe patterns can be done by applying wavelet filters [Kru99, Kru01].

The variations coming from the object reflectivity may occur with high spatial frequencies and cannot be filtered out by any spatial frequency filter. Similarly, the diffraction patterns lie in the same frequency bands as the desired interference pattern and cannot be filtered out. The electronic noise is a random fluctuation in time and can be avoided by averaging over a sequence of interferograms recorded at different time [Lut89].

The object, which is illuminated, is never ideally flat. The illuminated surface points emit spherical waves, which interfere and result into the speckle pattern. The speckle form a random

1 Introduction

pattern in space, which is stationary in time and disturb the interference fringes in the interferogram, decreasing thus the resolution and accuracy of the experiment.

1.4.7 Sign ambiguity, 2π-uncertainty

In the interferogram the $\cos[\phi(x, y)]$ instead of $\phi(x, y)$ is recorded. The extraction of the phase $\phi(x, y)$ from interferogram is assigned with a problem of the sign ambiguity and 2π-uncertainty of the cosine function:

$$\cos(\phi) = \cos(\pm\phi + 2\pi n), \quad n \in Z \tag{1.13}$$

The consequence of Eq. 1.13 is that it is not possible to say if the phase increases or decreases and that the value of ϕ can be determined just by modulo 2π (i.e. within the interval $(-\pi, \pi)$). The modulo 2π effect is shown in Fig. 1.6 for easier understanding. In the upper part several phases ϕ are shown that lead to the same function $\cos(\phi)$ depicted in the lower part. The modulo 2π effects are corrected by the processing step called demodulation, continuation or phase unwrapping, which will be discussed in the Chapter 1.5.

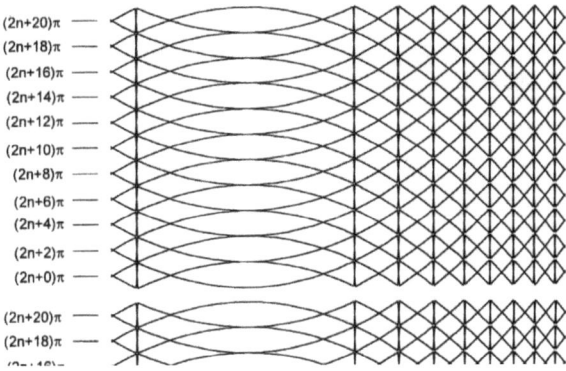

Fig. 1.6. Phase ambiguity (after [Kre96])

1 Introduction

The sign ambiguity can by avoided by doing an additional experiment [Jup88] or using of an additional information about the experimental conditions. This information can be e.g. a knowledge, if the plasma-optical or thermo-optical effect is dominant, or by recording two interferograms with defined shift of the object, or by introducing a carrier frequency for the fringes [Kre96].

1.4.8 Interferogram evaluation methods – overview

If the recording medium is e.g. a digital camera, the interferogram is digitised into an array of finite number of pixels. The $I(x, y)$ in Eq. 1.11 is quantised into finite number of grey levels, where (x, y) denote the pixel coordinates. The goal of the interferogram evaluation is to extract the phase distribution $\phi(x, y)$, which holds information about the refractive index change Δn. For this various phase extraction methods have been developed. They differ in the approach and experiment requirements. The optimal method is chosen according to the particular experiment facility and conditions.

The methods can be divided into four basic categories: *fringe skeletonizing*, *temporal heterodyning*, *phase sampling* and *Fourier transform* evaluation. The methods overview is well done in [Kre96] in Chapter *Quantitative evaluation of the interference phase* and in [Rob93, Ras94, Kuj98].

1.4.8.1 Fringe skeletonizing

Fringe skeletonizing methods are based on fringe counting. The fringe local maximum and minimum are located and the phase distribution is obtained by interpolation of the skeleton lines. The processing scheme consists of few steps [Kre96, Ras94, Kud95, Kre91].

First step is the signal-to-noise (S/N) ratio improvement (filtering by e.g. low-pass filter, median filter, Wiener filter [Lim90]) and shading correction (low-pass filter, spatial filter, background subtraction, Gaussian fitting). This is necessary for the local extreme localisation.

Second step is the fringe skeleton extraction. It is realised by segmentation, fringe tracking or other techniques. In segmentation technique the interference pattern is segmented into regions representing ridges, valleys and slopes of the fringes. The skeleton is obtained by finding the centers of the regions. In the fringe-tracking algorithm first a line, which cuts the

1 Introduction

fringes, is done and local maximum (minimum) are selected as the starting points. The tracking then follows the ridges (valleys) characterised by local intensity maximum (minimum) or by a derivative of grey levels. Another *skeletonizing* method is called phase-locked loop. The reference mirror is piezo-electrically oscillating in a short range (below the wavelength) with certain frequency. A bandpass filtering, centered on this frequency, yields to a signal, which is zero in the fringe extreme. Thus the skeleton is derived.

After the skeleton lines are found, they are numbered. Finally the interference phase have to be interpolated, according to the values of the skeleton lines (e.g. by the spline method, bilinear interpolation or interpolation by triangulation).

The main disadvantage of this concept is that it is sensitive to errors in the interferogram, noise and background correction. Any line intersection, merging or ending, missing points and artifacts report an error. In addition, the exact phase is extracted only at the fringe extreme and the phase in-between is obtained by interpolation.

1.4.8.2 Temporal heterodyning

Temporal heterodyning means the interference of two waves of slightly different frequency [Kre96, Ras94, Mas79]. The two reference beam method in conjunction with the double time exposure holography is the standard approach. By double time exposure holography the two wave fields have to be recorded and reconstructed with two reference beam holography setup. The frequency of the two reference beams is shifted e.g. by acousto-optic modulator. The interference pattern reconstructed with two reference waves is then time dependent. For its detection a fast two-dimensional detector is necessary (which is not still on the market) or a point sensor have to scan over the reconstructed image.

Temporal heterodyning is widely used in interferometric length measurement, but has only a limited application in the holographic interferometry. This is mainly due to necessity of using of high bandwidth detector, which is only a point detector and due to necessity of a high mechanical stability of the setup during recording and scanning, especially due to long lasting scanning. More details can be found in [Kre96].

1.4.8.3 Phase sampling

An alternative to the *temporal heterodyning* is the *phase sampling* [Kre96, Ras94, Vro91, Rob93, Kru97, Qua02, Kre91]. The frequency shift in one of the interfering waves of the

1 Introduction

heterodyne method is substituted by a very slow continuous or stepwise phase shifting between the interfering waves. Sometimes these methods are called quasi-heterodyne methods. The intensity distribution (see Eq. 1.11) recorded in this way is expressed by *phase sampling* equation:

$$I_n(x, y) = A(x, y) + B(x, y)\cos[\phi(x, y) + \phi_{Rn}] \qquad n \in N \qquad (1.14)$$

where ϕ_{Rn} is the shifted reference phase belonging to the *n*-th intensity distribution I_n. The relative phase shift ϕ_{Rn} can be achieved by using of two-reference-beam holographic setup, where the two reference beams are relatively shifted. This allows simultaneous recording of two interferograms. Other option is to place a reference mirror on a piezo crystal to shift the mirror.

In the case, when the reference mirror moves continually, the intensity is integrated during the recording time. The phase changes linearly. This is called phase shifting. The method, when the phase is shifted in fixed steps between two exposures and kept constant during recording, is called phase stepping.

In practice, the phase steps $\phi_{Rn}-\phi_{Rn-1}$ between two interferograms I_n and I_{n-1} are usually constant and the delay between their recording is as short as possible, to minimize the vibration influence. In that case the non-linear system of Eqs. 1.14 is obtained. The equations can be rewritten to a system of linear equations for $\sin(\phi)$ and $\cos(\phi)$ and the wrapped phase ϕ can be simply calculated [Kre96]. The phase is then determined directly from the recorded interferograms, which is the main advantage of this procedure.

The phase step and phase shift methods, which record and evaluate a set of interferograms, are widely used because of their easy automation [Aim00, Bre86]. The interference pattern is calculated with a high accuracy at all pixels of the interferogram, even without sign ambiguity. These methods are even insensitive to the dark areas. The only disadvantage is the necessity to record several interferograms of the object with a constant light intensity.

1.4.8.4 Fourier transform evaluation

The next method for the interferogram interpretation is the *Fourier transform* evaluation technique [Kre96, Ras94, Tak82, Rob93, Bon86, Kre86, Kre88, Kre91]. To explain the *Fourier transform* evaluation the interferogram given by Eq. 1.11 has to be rewritten using complex exponential:

1 Introduction

where:
$$I(x,y) = A(x,y) + C(x,y) + C^*(x,y) \qquad (1.15)$$

$$C(x,y) = \frac{1}{2}B(x,y)e^{i\phi(x,y)} \qquad (1.16)$$

where $i = \sqrt{-1}$ is the imaginary unit and * the complex conjugation. The two-dimensional Fast Fourier transform (FFT) applied to Eq. 1.15 yields to the spectrum function:

$$i(u,v) = a(u,v) + c(u,v) + c^*(u,v) \qquad (1.17)$$

where (u, v) are the spatial frequency coordinates. The spectrum $i(u, v)$ is a Hermitean distribution, since function $I(x, y)$ is real. It means, that $i(u, v) = i^*(-u, -v)$, i.e. the real part of spectrum is an even function, the imaginary part of spectrum is an odd function and the amplitude of the spectrum $|i(u, v)|$ is point-symmetric with respect to point (0, 0). Function $a(u, v)$ contains the zero peak $i(0, 0)$ and low frequency variations of the background. Functions $c(u, v)$ and $c^*(u, v)$ contain the same information about phase shift $\phi(x, y)$ and are placed symmetrically with respect to point (0, 0).

The basic idea of the *Fourier transform* evaluation is the elimination of the spatial frequency terms $a(u, v)$ and $c^*(u, v)$. This is performed by bandpass filtering. The inverse Fourier transform applied to the remaining part $c(u, v)$ results into a complex function $C(x, y)$ defined in Eq. 1.16. From this the phase shift ϕ can be extracted:

$$\phi(x,y) = \arctan\frac{\operatorname{Im}(C(x,y))}{\operatorname{Re}(C(x,y))} \qquad (1.18)$$

As in the case of *temporal heterodyning* and phase sampling, this phase is a wrapped phase and additional phase unwrapping process is required.

Spectrum bandpass filtering

Filtering of the frequencies $a(u, v)$ and $c^*(u, v)$ is not a trivial task [Kre96]. Usually it is not clear whether the point (u, v) belongs to the spectrum component $c(u, v)$ or $c^*(u, v)$, or is it a

1 Introduction

combination of both. In such case the easiest way is to use a half plane filter, which eliminates any two neighbouring quadrants of the spectrum, see Fig. 1.7. By this the spectrum becomes non-Hermitean. In addition, the low frequency part of the spectrum defined by a lower cut-off frequency (u_{min}, v_{min}) is filtered out to minimize the effect of background variations and the high frequency part of the spectrum defined by an upper cut-off frequency (u_{max}, v_{max}) is filtered out to minimize the random noise.

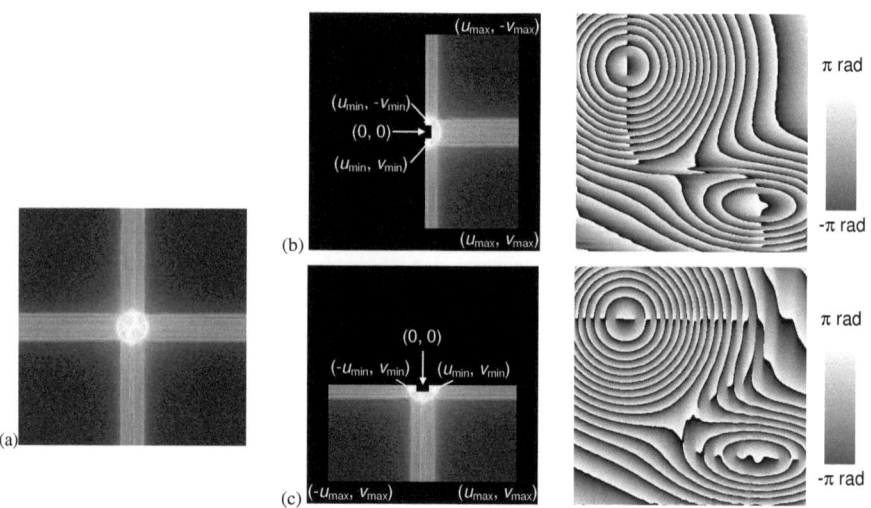

Fig. 1.7. (a) Spectrum of the interferogram from Fig. 1.5a. (b) +u-half-plane spectrum filter and calculated wrapped phase ϕ. (c) +v-half-plane filter and calculated wrapped phase ϕ.

Usage of the half-plane filter has a disadvantage. If for example a bandpass is the +u-half-plane (Fig. 1.7b), where only positive spatial frequencies in the horizontal (u) direction and both positive and negative frequencies in the vertical (v) direction can pass, the phase distribution with an increasing phase in the horizontal direction, but increasing and decreasing phase in the vertical direction is obtained (see Fig. 1.7b on the right). The information about the phase decreasing in the horizontal direction is missing. Phase extracted by using such a filter has to be combined with a phase extracted using an orthogonally oriented bandpass filter, see Fig. 1.7c. This can eliminate the phase error.

Fourier transform evaluation can be combined with phase step method. This combination eliminates problem of using the two half-plane filters and problem of the global sign ambiguity.

1 Introduction

The second interferogram has to be recorded with an additional phase step, best in the range between π/3 and 2π/3. Exact value of the step does not have to be known. Both interferograms are processed with the same bandpass filter and a sign function is determined from their combination. This sign function is then applied to the interferogram phase and the phase sign error is eliminated.

Spatial heterodyning

In spatial heterodyning an additional carrier frequency is added to the interference pattern, see Fig. 1.8a [Kre96]. This is done usually by tilting of the reference mirror or the object. The goal is to get equidistant linear fringes. This method requires a linear detector with high enough spatial resolution and uniform sensitivity. Let's assume that the spatial carrier frequency is (f_{Fx}, f_{Fy}). Eq. 1.11 can be rewritten to:

$$I(x,y) = A(x,y) + B(x,y)\cos[\phi(x,y) + 2\pi f_{Fx} x + 2\pi f_{Fy} y] \tag{1.19}$$

and its Fourier transform is:

$$i(u,v) = a(u,v) + c(u - f_{Fx}, v - f_{Fy}) + c*(u + f_{Fx}, v + f_{Fy}) \tag{1.20}$$

The spectrum components c and c^* are moved symmetrically in the spectrum domain and are centered around the points (f_{Fx}, f_{Fy}) and ($-f_{Fx}, -f_{Fy}$), respectively, see Fig. 1.8b. The component a remains centered around (0, 0). If the carrier frequency is high enough, the spectrum components a, c and c^* are well separated. This makes filtering out of components a and c^* much more easy and precise. The remaining component $c(u-f_{Fx}, v-f_{Fy})$ is shifted by vector ($-f_{Fx}, -f_{Fy}$) to the origin and thus function $c(u, v)$ is obtained. From this the wrapped phase $\phi(x, y)$ is calculated by inverse Fourier transform and Eq. 1.18.

Spatial heterodyning has been used for recording of fast events in [Liu02], where three frames with resolution 6 ns and frame interval 12 ns were recorded in a single CCD frame by using three different carrier frequencies. Each individual frame is reconstructed by digital filtering in the spectrum domain.

1 Introduction

Fig. 1.8. (a) An interferogram corresponding to the phase ϕ shown in Fig. 1.5b with introduced carrier frequency and (b) its spectrum. In the spectrum the spectrum components c, c^* are well separated. The component a is located only in the spectrum center [0, 0], since there are no background variations.

An analogue of the spatial heterodyning is a method called the spatial synchronous detection. Here the interferogram with the fringe carrier frequency (f_{Fx}, f_{Fy}) is multiplied by $\cos(2\pi f_{Fx}x+2\pi f_{Fy}y)$ and by $\sin(2\pi f_{Fx}x+2\pi f_{Fy}y)$. Thus the component a is shifted in the spectrum domain by a vector (f_{Fx}, f_{Fy}) and the c component is localised around (0, 0). After applying a low-pass filter to the spectrum of both results, the phase is calculated by Eq. 1.18.

1.4.9 Comparison of phase extraction methods

The advantages ("+") and disadvantages ("-") of above described phase extraction methods are summarised in Table 1.1. Precision of all the methods is limited by insufficient quantization, spurious diffraction or reflections, aberrations of the optics, vibrations, air turbulence, inhomogeneity of reference beam wavefront, detector nonlinearity etc. A detailed comparison of fringe pattern analysis methods is done in [Kuj98].

Table 1.1. Comparison of phase extraction methods.

Category	Method	+/−	
Skeletonizing	Fringe tracking	− exact phase calculated only at fringe extreme − inaccurate interpolation between skeleton lines − sign ambiguity − high sensitivity to noise and background variations	
	Segmentation		
	Phase-locked loop		
Temporal heterodyning		− either need of 2D-sensor with high bandwidth or precisely moving point sensor − difficult setup arrangement + high accuracy, automated, environment insensitive	
Phase sampling	Phase shifting	Commercially used method for static objects − at least 3 object interferograms necessary − performing of phase shift in the setup − sensitive to noise, detector nonlinearity, light intensity stability and environment stability	
	Phase stepping	+ fully automated + phase calculated in all pixels, without sign ambiguity + indifferent to black margins	
Fourier transform evaluation	Using of two spatial filters		− 2D spatial filtering − global sign ambiguity + noise and intensity variation filtering + usually a single interferogram sufficient
	Phase step method	− need for 2 object interferograms	
	Spatial heterodyning	Commercially used medium accuracy methods for dynamic phenomena	
	Spatial synchronous detection		

1.5 Phase unwrapping

As was described in Chapter 1.4.5, the extracted phase calculated by any of the method has values laying between $-\pi$ and $+\pi$ (see Figs. 1.7b,c), due to the 2π-uncertainty of cosine function (see Eq. 1.13). In standard applications the phase demodulation is required to change the "saw-tooth" phase shape into a continuous phase distribution. The phase 2π-jumps are eliminated by addition or subtraction of 2π multiples to every point of the interferogram. Correct

1 Introduction

addition (subtraction) can be done if exact phase in one point of the interferogram is known. The phase of the whole interferogram is then corrected according to this point.

The unwrapping process is depicted in Fig. 1.9. In Fig. 1.9a the wrapped phase is shown. The corresponding step function shown in Fig. 1.9b must be found and added to the wrapped phase. This is the main task of the unwrapping. The addition results into the continuous phase in Fig. 1.9c. If the step function is marked like $n(x_i)$ in pixel x_i, the continuous phase shift is then given by a sum of the step function and the wrapped phase $\phi(x_i)$:

$$\phi_{contin}(x_i) = \phi(x_i) + n(x_i) \tag{1.21}$$

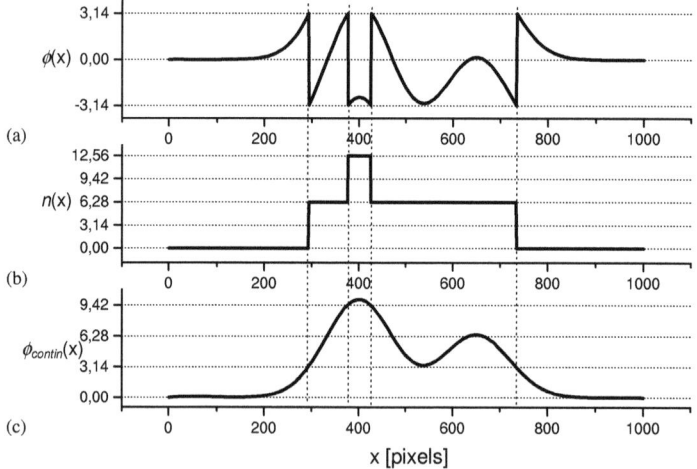

Fig. 1.9. Principle of phase unwrapping. (a) Wrapped phase, (b) step function, (c) unwrapped phase.

The unwrapping methods can be generally divided into two categories: path dependent techniques and path independent techniques. In the path dependent techniques the order in which the pixels are processed is predetermined by the technique. In the path independent techniques the order is determined by the phase values at the pixels. The most important methods are described bellow.

1 Introduction

1.5.1 *Straightforward* algorithm

Straightforward algorithm [Kre96, Vro91, Rob93, Ras94] belongs to the path dependent techniques. The unwrapping in one line (row or column) of the interferogram is done by comparison of the phase difference between two neighbouring pixels. The step function $n(x_i)$ in pixel x_1 is set e.g. to zero. The step function in pixel x_i is created as following [Kre86, Vro91]:

$$n(x_i) = \begin{cases} 2\pi n(x_{i-1}) & \text{if } |\phi(x_i) - \phi(x_{i-1})| < \beta\pi \\ 2\pi n(x_{i-1}) + 2\pi & \text{if } \phi(x_i) - \phi(x_{i-1}) \leq -\beta\pi \\ 2\pi n(x_{i-1}) - 2\pi & \text{if } \phi(x_i) - \phi(x_{i-1}) \geq \beta\pi \end{cases} \qquad i = 2,3,4..., \qquad (1.22)$$

where $\beta \in \langle 0, 2 \rangle$ is a tolerance factor, usually equal to 1. If β was equal to 2, all pixels are assumed to have an accurate value.

In the case of two-dimensional phase, the step function for e.g. one column is calculated according to Eq. 1.22. The pixels in this column act as starting pixels for row unwrapping, see Fig. 1.10a. Another trivial variations are possible. The two-dimensional data are treated like a set of one-dimensional data.

The number of the error step recognition is decreased, if the phase differences to the pixel above in the previous row and the one to the left in the same row are checked. If both differences indicate the same step function, this is taken for unwrapping. If differences disagree, the pixel is unwrapped later.

This algorithm is simple to implement but it is strongly sensitive to the phase errors (e.g. speckles). If a wrong step function is calculated due to the noise, this error spreads up to the last pixel in the row (column).

1.5.2 *Spiral scanning* algorithm

In *spiral scanning* algorithm [Kre96, Vro91, Rob93, Ras94], the starting point is usually the center point of the two-dimensional phase, see Fig. 1.10b. The current pixel value is compared to the mean of a set of previously unwrapped pixels in 3x3 neighbourhood. Pixels, which have not been unwrapped yet, are ignored. The data is scanned spirally in order to

1 Introduction

maximise the number of already unwrapped pixels in the neighbourhood. Thus in most cases 4 unwrapped pixels are present in this neighbourhood, but 3 are usually sufficient.

The same restriction as for the *straightforward* algorithm is valid and this is, that all pixels have to be valid pixels, no masking of invalid pixel is allowed [Vro91].

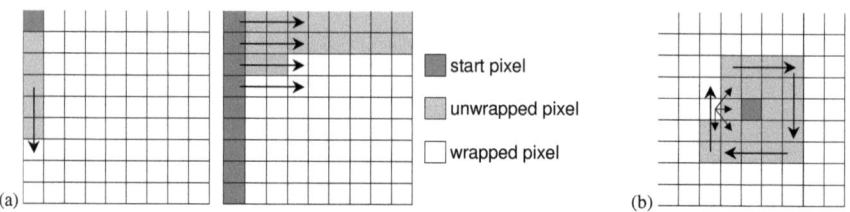

Fig. 1.10. Phase unwrapping by (a) the *straightforward* algorithm, (b) the *spiral scanning* algorithm (after [Vro91])

1.5.3 *Pixel queue* algorithm

In the *pixel queue* algorithm [Vro91, Rob93] the data are scanned like a fluid spreading over the object but around invalid pixels, i.e. masked defects. For this, masking of the invalid pixels has to be done first. Then a starting point is chosen, best in the middle of the area to be unwrapped. A pixel queue is a one-dimensional data array, to which the pixel addresses are stored on one side and fetched from the other side, according to the flowchart in Fig. 1.11a. The process is finished when the queue is empty.

Due to the fact, that 4 pixels sharing an edge with the current pixel are put to the queue, the processing propagates in a diamond shape. Other options for the queue filling up are possible and the shape of the propagation is then different, e.g. pixels in 3x3 neighbourhood can be stored to the queue [Vro91].

The disadvantage of this method is that two areas, which are completely separated by the mask, have to be processed separately.

1.5.4 *Minimum spanning tree* algorithm

Minimum spanning tree algorithm [Kre96, Rob93, Chi92] belongs to the path independent unwrapping techniques. This method minimises spreading of erroneous phase. Connections between two neighbouring pixels are called arcs. The value of the arc is given by $\min\{|\phi(x_1)-\phi(x_2)|, |\phi(x_1)-\phi(x_2)+2\pi|, |\phi(x_1)-\phi(x_2)-2\pi|\}$. The arc value interprets the pixels "confidence degree". The unwrapping sequence is shown in flowchart in Fig. 1.11b. The path is not given in advance but the unwrapping is done in direction, where the error in the phase is least probable. The pixels with the highest arcs are unwrapped at the end. This algorithm can be modified for less computation effort, if only the arcs with values less than some threshold are recorded to the list.

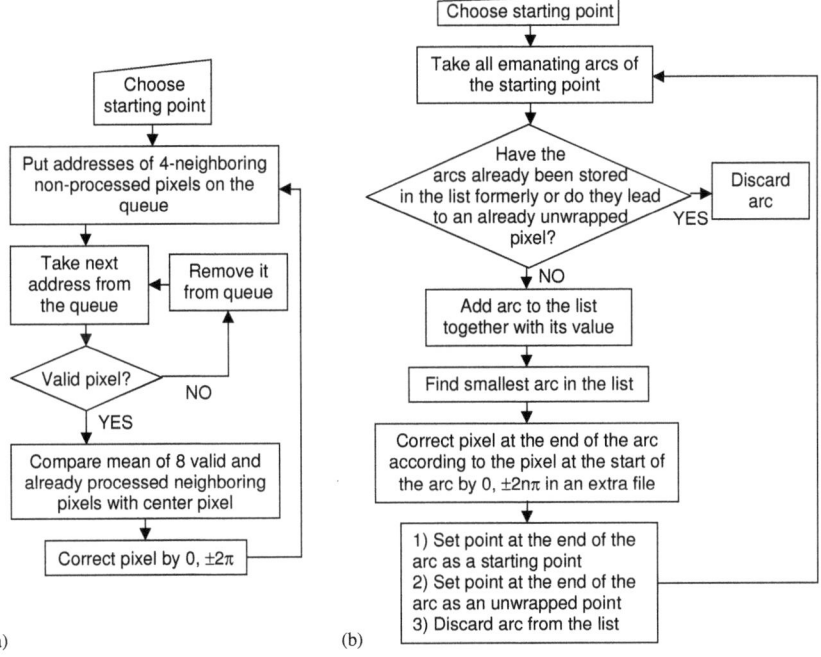

(a) (b)

Fig. 1.11. Flowchart of the (a) *pixel queue* unwrapping algorithm, (b) *minimum spanning tree* algorithm.

1 Introduction

1.5.5 *Cellular automata*

The *cellular automata* algorithm [Kre96, Rob93] is based on the effect of large number of pixels. It has two steps. In the first step called local iteration, the 0 or -2π or +2π is added to every point of the phase. To decide this, 4 neighbouring pixels are taken into account. After number of local iterations, oscillation between two patterns occurs. Then the arithmetic average of these two states is computed. This is called global iteration. The two steps are repeated, until a steady state is reached.

This algorithm is robust against distortions and noise. But there is also a possibility that it will never stop due to some local phase errors, or it can be very time consuming due to large number of pixels.

1.5.6 Other algorithms and approaches

Slightly different approach to the *minimum spanning tree* has been proposed. The arcs of the current pixel are calculated and the unwrapping is done in the direction of the smallest one [Rob93, Kwo87]. The process repeats for the newly reached pixel.

Radically different approach is called a bandlimit unwrapping [Kre96, Rob93]. Here all possible step functions are tested and the spectrum bandwidth is checked. The step function, which minimised the bandwidth, is selected. This is because the sharp phase steps make the spectrum broader. The method can fail in the case of large number of fringes.

To avoid spreading of the error to the rest of the two-dimensional phase, the phase can be divided into rectangular tiles [Kre96, Rob93]. In each tile the phase is unwrapped by one of the unwrapping method mentioned above. Then the edges of the tiles are compared and adjusted by adding (subtracting) 2π for the whole tile.

A similar approach is to divide the phase into regions without phase errors. After local demodulation, the regions are phase shifted in order to minimize the discontinuities between them.

Another unwrapping technique is proposed in [Bon91], the noise immune algorithms are in [Kad97, Hua02].

1 Introduction

1.5.7 Comparison of phase unwrapping methods

The summary of the advantages ("+") and disadvantages ("-") of above described phase unwrapping methods is in Table 1.2.

Table 1.2. Comparison of phase unwrapping methods.

Phase unwrapping method	+/-
Straightforward algorithm	+ simple to program
	+ high speed
	- sensitive to a single pixel phase defect
	- erroneous step identification infect rest of the image
	- invalid pixel masking not allowed
Spiral scanning algorithm	+ simple to program
	+ high speed
	- sensitive to phase defect, but less that *straightforward* method
	- erroneous step identification infect rest of the image
	- invalid pixel masking not allowed
Pixel queue algorithm	+ masking of invalid data allowed
	+ spreads around masked pixels
	+ unwraps all opened defects
	- masked pixels are not demodulated
	- method can fail in case of large amount of invalid pixels
Minimum spanning tree algorithm	+ unwrapping path adapts to the current phase
	+ pixels with low error probability are evaluated first and invalid pixels are unwrapped at the end
	+ all pixels are processed
	- complicated programming
	- long computation time
	- method can fail in case of large amount of invalid pixels
Cellular automata	+ robust against distortions and noise
	- long computation time
	- method can fail in case of very noisy data

1 Introduction

1.5.8 Error sources in the unwrapping process

During the phase unwrapping the 2π-steps are localised. The correct functionality of the unwrapping techniques is influenced by many effects. Different unwrapping techniques have different sensitivity to these effects. Incorrect phase step detection leads to a wrong phase interpretation and the error can spread over the image, sometimes resulting to a totally useless result. Therefore the optimal unwrapping procedure has to be chosen according to the particular case and the sources of the defects have to be reduced.

The first important prerequisite for correct phase unwrapping is a sufficient interferogram sampling [Rob93]. If the data are undersampled (necessary at least 2-3 pixels per fringe), it can lead to a small phase difference between two neighbouring pixels, where a phase step occurs. The phase step is then not detected. If the data are oversampled, some unwrapping procedures can fail due to introduction of non-existing phase steps.

The statistical noise (e.g. electronics noise, speckle noise) in the interferogram is during the phase extraction process transferred to the wrapped phase. If the amplitude of the noise approaches 2π, the actual phase step becomes obscured [Rob93]. This effect is reduced using the median filter for example, which preserves the 2π-steps. The smoothing by arithmetic averaging is not permitted since it smoothes the sharp 2π-steps [Rob93, Vro91]. Regions of poor contrast of the interference pattern can lead to many errors located in a small area. These regions have to be flagged or masked before starting the unwrapping [Rob93].

Objects with sharp edges are another source of errors, since it results into a dislocation in a fringe and consequently into discontinuities in the phase [Rob93]. The regions of discontinuities must be masked during unwrapping to avoid wrong phase interpretation and error spreading. If the object is not topologically connected to the rest of the area, the phase can be unwrapped separately within the area of the object. If there is an external information about the object height, the phase between the unconnected areas can be derived, as well as the absolute phase value.

The phase errors can be divided into three classes [Rob93]. A simple defect is when the valid data around the defect can be accessed from a number of directions, see Fig. 1.12a. Then e.g. unwrapping from top left to bottom right can process all valid data by counting around the defect. If the boundary of the defect is opened (Fig. 1.12b), then to unwrap all the valid phase data requires a complex scanning path chosen for this particular data set or one that can adapt to

1 Introduction

the data itself. These defects are e.g. in the shape of character C. If the boundary of the defect is closed (Fig. 1.12c), i.e. some valid data are inside the defected area and not accessible from any direction, then these data can be unwrapped only independently without relation to other areas.

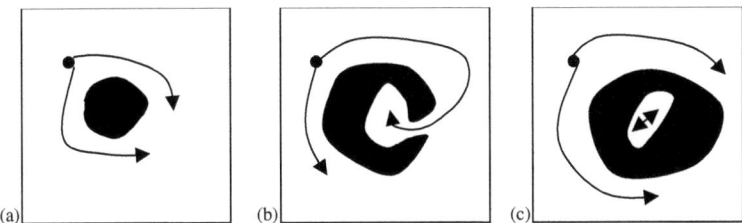

Fig. 1.12. Classification of phase defects: (a) simple defect, (b) opened defect, (c) closed defect (after [Rob93])

In case of two-dimensional data the phase jumps can be detected by scanning through the phase data in different directions. The integral of the phase along any path that starts in point 'P' and ends again in the same point 'P' is constant and independent on the path [Rob93]. This is the advantage of the two-dimensional data and can be used to check the phase in any point. However, there is a large number of paths leading back to the same point.

1.6 2D power dissipation density

The phase shift $\varphi(x, y, t)$ reflects the total history of the spatial and temporal dependence of the power dissipated in the device. It gives not an information about the instantaneous state of the device. This information is located in the instantaneous 2D power dissipation density P_{2D}. The P_{2D} represents the current flows at certain time instant. This is an important information for the study of the dynamics of moving current filaments in the devices.

The P_{2D} is an integral of the three dimensional power dissipation density P_{3D} along the laser beam path:

$$P_{2D}(x,y,t) = \int_0^L P_{3D}(x,y,z,t)dz = \frac{\partial E_{2D}(x,y,t)}{\partial t}, \qquad (1.23)$$

1 Introduction

where L is the substrate thickness and E_{2D} is the 2D energy density. In case when the free carrier effect can be neglected, the temperature induced phase shift distribution $\varphi(x, y, t)$ in the device at a particular time instant t is proportional to E_{2D} in the device [Pog02c]. The P_{2D} can be measured e.g. with the heterodyne setup [Pog02a, Pog03c], which is however time consuming. Neglecting the heat transfer to the top device layers (normal component of the heat flow vector), P_{2D} is given by [Pog02a]:

$$P_{2D}(x, y, t) = \lambda \left(4\pi \frac{dn}{dT} \right)^{-1} (\Sigma - \Psi), \qquad (1.24)$$

where:

$$\Sigma = c_V \frac{\partial \varphi(x, y, t)}{\partial t}$$

$$\Psi = \kappa \frac{\partial^2 \varphi(x, y, t)}{\partial x^2} + \kappa \frac{\partial^2 \varphi(x, y, t)}{\partial y^2}$$

where dn/dT is the temperature coefficient of the refractive index (1.8×10^{-4} K^{-1} for Si), c_V is the volume specific heat (1.631×10^6 JK^{-1}m^{-3} for Si), $\varphi(x, y, t)$ is the measured phase shift and κ is the thermal conductivity (vary from 150 to 50 WK^{-1}m^{-1} for temperature range from 300 to 600 K for Si). Thus the P_{2D} is calculated from the time Σ and spatial Ψ derivatives of the optical phase shift.

2 2D setup

In this part of the thesis an optical setup for the nanosecond imaging of the refractive index changes in the DUT is introduced. The main task of this work was to develop and optimise the configuration and the optics of a laboratory setup, to extend setup facility to measure devices of size from few micrometers to several millimeters, to analyse the errors in the interferogram, to find an optimal phase extraction technique for this application of the interferometry and to automate the phase extraction process. Based on this knowledge finally a probing system for the industrial measurement on the wafer level and measurement of flip-chip packages was developed.

The setup hardware is based on a Michelson interferometer where the refractive index is imaged using a 2D holographic interferometry (see Chapter 1.4.5). A pulsed laser source is used in order to create an interferogram of the DUT corresponding to certain time instant during the stress pulse. Any optical path difference is then obtained by comparison of the interferograms recorded before (reference interferogram) and during the stress pulse (stressed interferogram). In this chapter the configuration of the setup for the measurement at one time instant and two time instants is shown and the setup parameters are resumed.

2.1 Setup description

2.1.1 Imaging at single time instant

The layout of the experimental setup for imaging at single time instant is shown in Fig. 2.1a and its photograph in Fig. 2.1c. Part of the laser beam is reflected by a glass plate to the optical detector (DET) before it enters the Michelson interferometer. This allows alignment the laser pulse position according the stress pulse on the oscilloscope. The non-polarizing beam splitter (NPBS1) splits the beam to the probe and reference branch of the interferometer. In the probe branch the beam is reflected from the DUT and in the reference branch the beam is

2 2D setup

reflected from a reference mirror. These two reflected beams are combined and projected by a lens L2 to the IR camera, where they create an interference pattern. All the optics has an antireflection coating for wavelength range 1050-1550 nm to minimize stray reflections. Frames from the camera are stored to the computer using IMAQ (image acquisition) card. The IR lamp allows acquisition of the backside IR images of the DUT. The DUT is mounted on a printing-board on x-y-z-Φ stage and contacted electrically by flexible cables (see Fig. 2.1d).

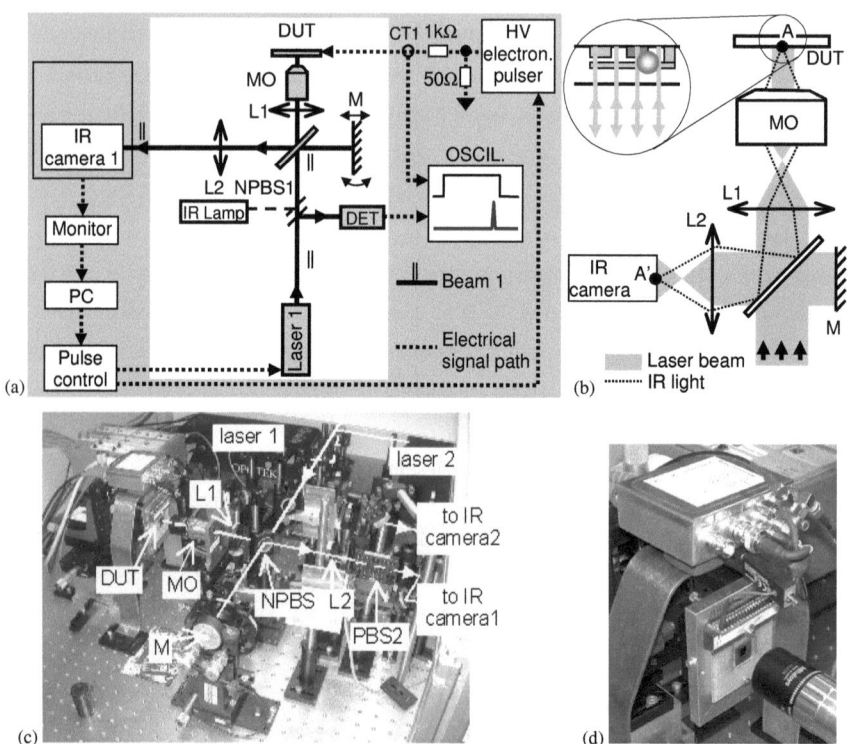

Fig. 2.1. (a) Schematic layout of the 2D TIM setup for one time instant measurements, (b) diagram showing the geometry of laser beam and IR light path, (c) picture of the setup and (d) detail of the DUT mounting and contacting.

The detailed beam path is shown in Fig. 2.1b. The multi-mode laser beam is broad with a low divergence (< 10 mrad) when entering the interferometer. In the probe branch, the beam passes a telescopic optical system composed by a lens (L1) and microscope objective (MO).

Thus the laser beam outgoing from the objective remains quasi parallel and illuminates the whole DUT. Due to this geometry the beam reflected from the DUT is transformed back by the MO and L1 and the interference with the reference beam results into parallel fringes.

Furthermore the optical system composed of MO, L1 and L2 serves as an image transformation system, which transforms the point A laying in the plane of the DUT into the point A' laying in the plane of the camera (see dotted lines in Fig. 2.1b). This condition is necessary to create an IR image of the DUT on the camera.

2.1.2 Imaging at two time instants

For detailed investigation of unrepeatable behaviour and of dynamics of the destructive processes the measurement of the refractive index changes at second time instant during a single stress pulse was introduced. Investigations at two time instants are necessary for understanding of the device internal behaviour and physical processes in the device. With the heterodyne technique or any other commercially used technique the unrepeatable processes can not be investigated since these techniques require repetitive pulsing. Furthermore with this option the P_{2D} in the device can be measured as will be described in Chapter 2.7. Such P_{2D} measurement is much more time-efficient than measurement with the heterodyne technique. To obtain two interferograms during a single stress pulse two variants of the setup were proposed and analysed, see Fig. 2.2.

2.1.2.1 Delay line variant

In the first layout (Fig. 2.2a), 50 % of the laser beam is deflected by a non-polarizing beam splitter NPBS2 to a delay line. The delay line is composed by a series of mirrors and focusing lenses to avoid energy loses due to laser beam divergence. The laser beam at the output of the laser is linearly polarised. The geometry of the mirrors in the delay branch was chosen in such way, that the plane of polarisation of the beam is 90 degree rotated. The polarizing beam splitter cube (PBS1) is then used to merge the two laser beams to the same interferometric setup. The advantage of such geometry is that there are no optical power loses during merging of the two beams. The length of the delay line defines the delay between the two laser pulses, which can be then seen on the oscilloscope. After the two laser beams pass the lens L2, they are split

(PBS2) according to their polarisation into two cameras. Thus every camera records the interferogram of thermal distribution in the DUT at different time instant.

The disadvantage of the layout in Fig. 2.2a is that the delay between the two laser pulses is fixed by the geometry of the delay line. It can be changed only by rearrangement of the delay branch, that is a time consuming process. From that reason two delay branches with different length have been constructed: a branch with length 6 m (20 ns delay) and a branch with length 9 m (30 ns delay). In the case of 20 ns delay the quality of the interferogram is comparable to the interferogram exposed with the non-delayed beam. In the case of 30 ns delay the amplitude of interference fringes has decreased, see Fig. 2.3. In some areas the fringe amplitude becomes too small for correct phase extraction. This is due to mixing of the spatial modes of the multimode laser beam, which leads to the decrease of the spatial coherence of the beam. For even longer delay branch the fringe amplitude decreases more and the phase shift can not be extracted.

2.1.2.2 Two lasers variant

Since in the experiments a longer delay is required and the delay branch rearrangement needs a lot of effort and time, a variant shown in Fig. 2.2b has been constructed. Here a second laser is used instead of the delay branch. The polarisation of this beam is perpendicular to the polarisation of the first laser beam. Therefore the two laser beams can be easily combined by the PBS1. The relative delay between the two laser beams is computer controlled and synchronised and it can vary from 0 ns to several seconds. The only disadvantage of this layout is the high price of the laser.

2 2D setup

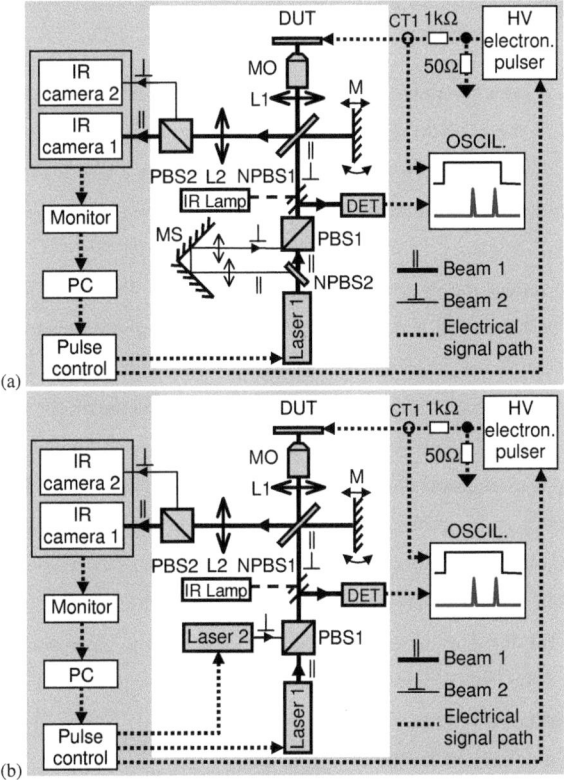

Fig. 2.2. Two realised layouts of the 2D TIM setup for measurement at two time instants. The delayed beam is realised (a) by a delay branch and (b) by a second (delayed) laser.

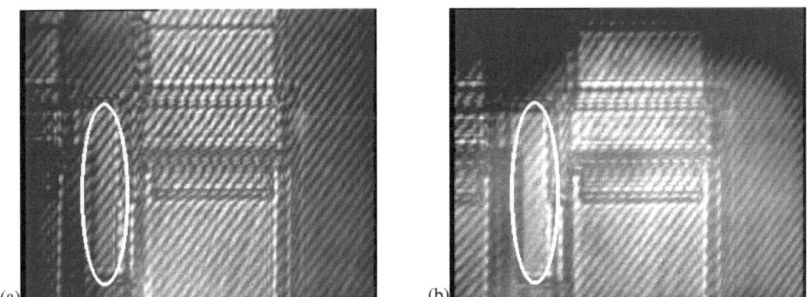

Fig. 2.3. Examples of the interferograms recorded in the setup in Fig. 2.2b (a) by the not-delayed laser beam and (b) by the beam delayed 30 ns. The laser beam passed lens 6 transitions and mirror 11 reflections in the delay branch. The fringe amplitude in the case of (b) has approx. 30 % decreased in average. In marked area the fringe contrast decreased to the level, when the phase extraction becomes difficult.

2.1.3 Probe station

In order to be able to contact the DUT by the needle probes and thus avoid gluing and bonding of the DUT, the 2D TIM setup was incorporated into a probe station SUSS MicroTec PM8DSP, see Fig. 2.4a. Here the wafer or the chip is placed on a kinetic chuck (see Fig. 2.4a) and contacted from the top by the needles. The device is inspected by the laser from the bottom side.

In order to mount the optics bellow the chuck and to illuminate the device from the bottom side, the probe station was redesigned together with the producer according to our requirements. The chuck was lifted up together with the needle probes and microscope. Here a compromise between the space for the optics below the chuck and the mechanical stability of the chuck was chosen. An oval hole was made in the chuck in order to illuminated the device from the bottom. The optical setup was designed to be as compact as possible, see Fig. 2.4b. A special adapter for the optics in the probe branch was manufactured (see Fig. 2.4b). The lasers and cameras are placed outside the probe station. A removable mirror reflects the laser beams either to the probe station or to the laboratory setup. To change the field of view to 5 mm two kinematic mount for the lens system and for the system with microscope objective were designed, see Fig. 2.4e. This requires minimum of adjustment when exchanged. On the probe station the measurement at two time instants is possible.

This setup was adapted for the use in the industry like environment. Therefore it allows analysis of DUT on the wafer level, see Fig. 2.4c. As well a special adapter was made for investigation of small wafer cut-outs, see Fig. 2.4d. The vacuum fixes the wafer or the sample adapter to the chuck. The minimum sample size that can be investigated here is 2x3 mm, the maximum diameter of the wafer is 300 mm. The contacting of the DUT by the needles is done manually and the user controls it trough the microscope mounted above the chuck.

2 2D setup

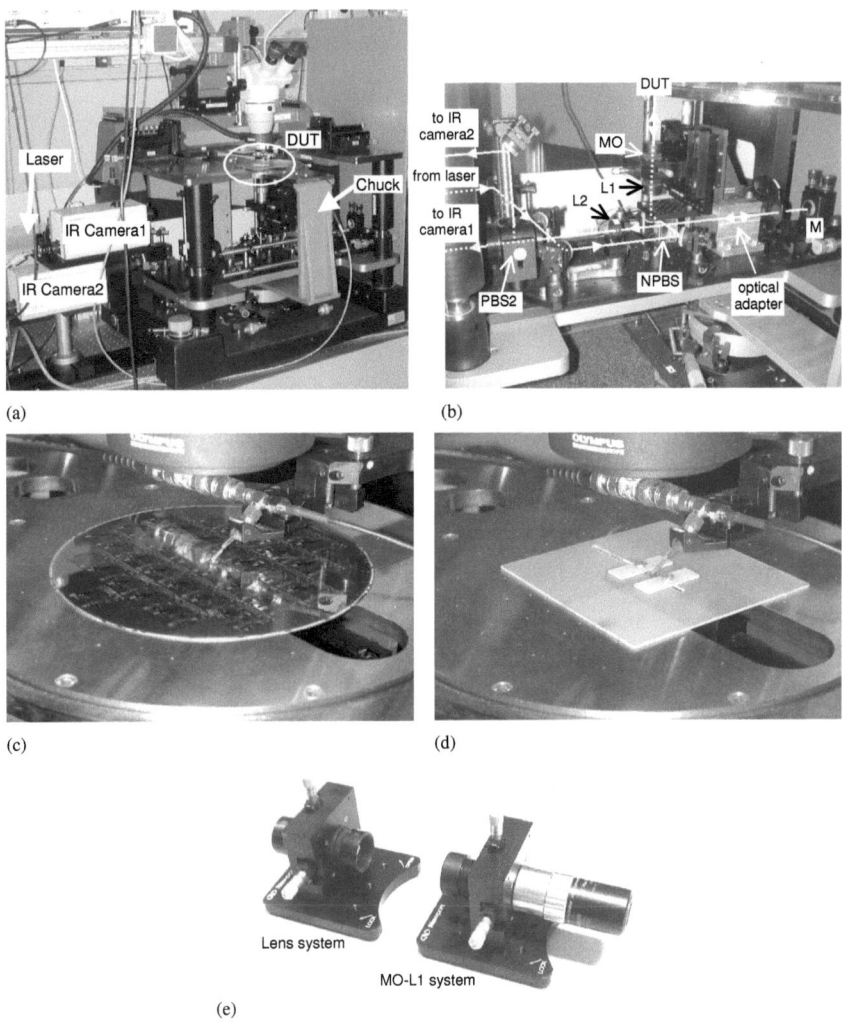

Fig. 2.4. 2D TIM setup implemented into a probe station. (a) General view, (b) details of the optics, (c) probing on the wafer level, (d) adapter for small wafer cut-outs, (e) kinematic mounts with optics for different fields of view.

2.2 Detection scheme

In this chapter the sequence of the measurement procedure, setup triggering, setup timing and the data flow is explained.

2.2.1 Measurement framework

The general flowchart of the measurement procedure in the 2D TIM setup is shown in Fig. 2.5. In Step 1 the sample is prepared for the measurement (gluing on the printing-board, bonding, mounting in the setup, contacting, setup alignment etc.). In Step 2 the measurement parameters like amplitude and duration of the stress pulse, number of stress pulses, number of reference interferograms, number of frames recorded per one stress pulse, optical pulse delays, parameters for the IV measurements, backside IR image acquisition, file names etc. are set in the acquisition software (programmed in *LabView* [Hee05]). During the measurement (Step 3) the interferograms and waveforms from the oscilloscope are recorded by the computer. The reference interferograms and the stressed interferograms are recorded fast after each other to minimize the influence of the vibrations and optical table instabilities. The last Step 4 is the interferogram processing, where the phase shift is extracted from the interferograms. For this, an analysis software has been developed (programmed in *Matlab*), see Appendix for details.

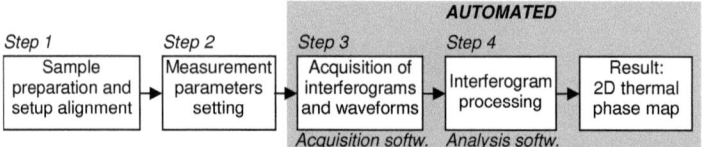

Fig. 2.5. Measurement procedure flowchart.

2.2.2 Triggering and data acquisition

The schematic structure of the instrument interconnection and the data acquisition is shown in Fig. 2.6. The figure shows the triggering signals, electrical and optical signals and digital data flow. The triggering and the data acquisition are fully computer controlled and automated. The computer generates 20 Hz TTL clock signal, which triggers the flash lamp of laser 2 (requirement of the laser producer) and the switch box. In switch box it is divided into 10 Hz and 1 Hz signal. The 10 Hz signal triggers the flash lamp of laser 1 (laser producer requires for the operation of the second laser 10 Hz repetition frequency) and the 1 Hz signal triggers the delay generator and the IMAQ card. The delay generator 1 controls the delay between the stress pulse and the first optical pulse. It has two outputs. One output triggers the stress pulse generator, second output controls the q-switch of the laser 1 and also triggers the second delay generator. The second delay generator controls the relative delay between the two laser pulses by controlling the q-switch of laser 2. After the computer gets the trigger pulse to the IMAQ card, the acquisition of the specified number of frames from the cameras starts and the data from the oscilloscope are downloaded.

The typical timing diagram of the used laser is shown in Fig. 2.7. The laser flashlamp and Q-switch is triggered by a TTL signal. From the flash lamp trigger to the moment of maximum neodymium fluorescence it takes approximately 190 μs. After this time the Q-switch has to be opened to ensure the optimal laser generation. It was experimentally found that the delay 190 μs can vary in the range ± 5 μs without effecting of the generation efficiency. For shorter or longer delays the laser output energy decreases.

The timing diagram of the complete 2D TIM setup is shown in Fig. 2.8. The figure is split into the left and right part. The timing diagram for the stress pulses shorter than 5 μs is on the left side, the timing diagram for the stress pulses longer than 5 μs and shorted than 100 ms is on the right side. The limit of 5 μs arises from the argument that the delay between the flashlamp and the Q-switch can vary only within 5 μs, as was explained above. Therefore for testing with stress pulses longer than 5 μs the timing has to be changed according to the diagram on the right side. Here the laser pulse has supplementary delay 100 ms, since this is the period between two flashlamp trigger events of laser 1. The stress pulse delay is then between 0 and 100 ms.

For stress pulses N-times longer than 100 ms the laser q-switch has to be delayed $N \times 100$ ms, where N is an integer.

2 2D setup

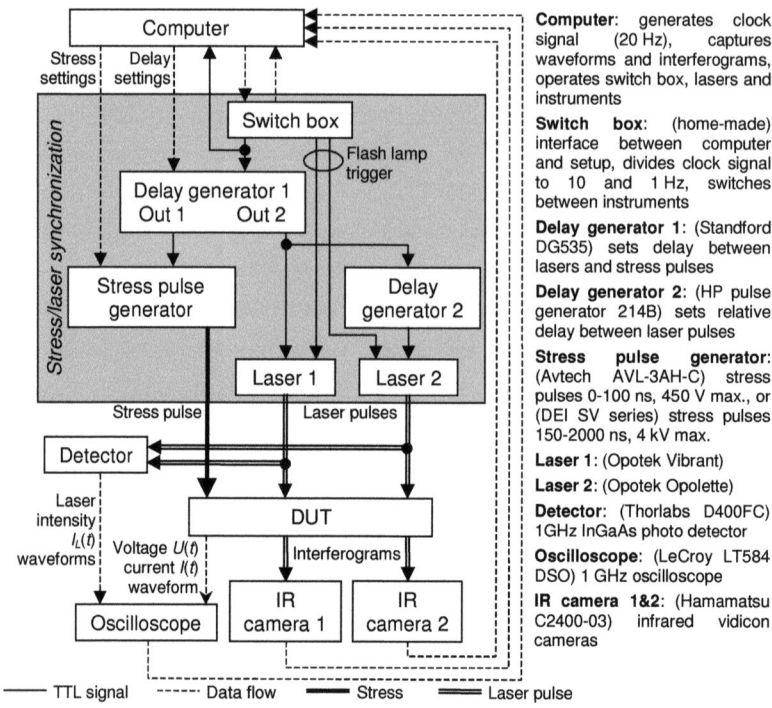

Fig. 2.6. Diagram of the instrument connection and the data flow.

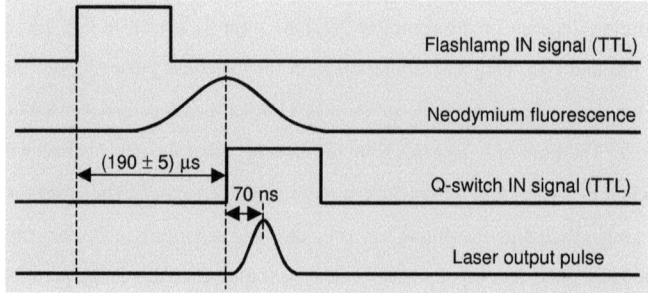

Fig. 2.7. Typical timing diagram of the pulsed laser Opolette 355 II.

2 2D setup

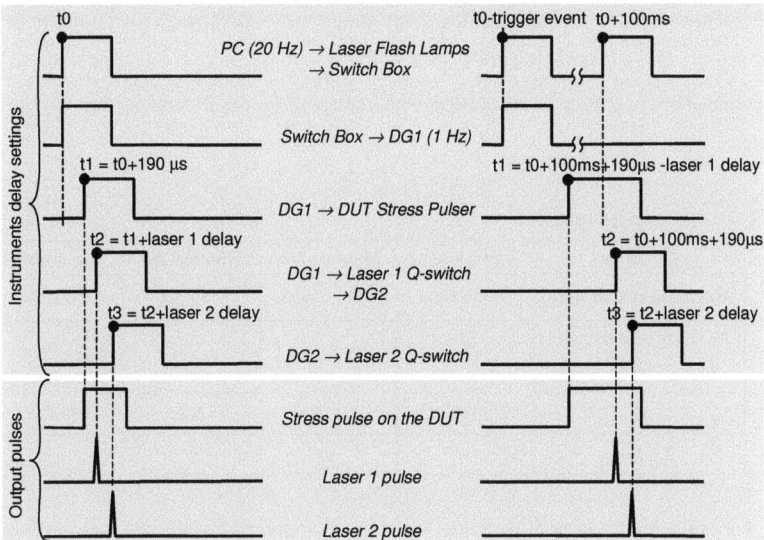

Fig. 2.8. 2D TIM setup timing diagram for the stress pulse < 5 μs (left) and > 5 μs (right).

2.3 Setup specifications

2.3.1 Field of view

The area of the inspected sample that is viewed on the camera will be called field of view. The maximal field of view is limited by the objective magnification and distribution of the light energy across the laser beam. By using the objectives with different magnification and by using a lens system (see Fig. 2.4e) the field of view is changed. The field of view has to be chosen in such way that the active area of the device (i.e. the heated area) does not cover the whole image. This is necessary for the phase extraction, since the exact value of the stress-induced phase shift can be obtained only with respect to the non-heated area (2π uncertainty, see Chapter 1.4.7)

The infrared microscope objectives used in the setup are manufactured by Mitutoyo. In Table 2.1 the field of view for the objectives that are used in the setup is given. Exchange of the

2 2D setup

microscope objectives is associated with a small correction of the distance between MO and lens L1, see Fig. 2.1. To reach the field of view 6 mm it is necessary to exchange the objective to a lens system. This exchange is associated with replacing of the kinematic mounts shown in Fig. 2.4e.

Table 2.1. 2D TIM setup specifications.

	Objective	Lens system	Mitutoyo 10x	Mitutoyo 20x	Mitutoyo 50x
1	Max. field of view (µm)	6000 x 6000	1500 x 1000	850 x 500	430 x 230
2	Pixel size in maximally zoomed image (µm)	15 x 15	1 x 0.9	0.6 x 0.5	0.3 x 0.3
3	N_A	0.02	0.26	0.4	0.42
4	d_{difr} (µm)	40	3	2	1.9

For optimal utilisation of the setup, the zooming to the area of interest was implemented. The zooming is done by changing the relative distance between the camera and lens L2 and between the MO and DUT, see Fig. 2.2. However, the zooming causes the decrease of the light energy density that comes to the camera and consequently the decrease of the interferogram contrast. Therefore the maximal available zoom is limited by the sensitivity of the camera and by the energy of the laser pulse. For the given microscope objective the maximal zooming factor is approx. 2.

2.3.2 Spatial resolution

There are two significant factors, which limit the spatial resolution of the experimental setup: size of the pixel in the interferogram image and diffraction limit of the optical system.

The analog signal from the IR camera is digitised by the IMAQ card to an array of pixels. The area of the DUT, which is imaged by a single pixel, will be called pixel size. Taking into account the field of view and the maximal zooming factor, the minimal pixel size can be calculated for every objective. The results are given in Table 2.1 in second row.

According to the Reyleigh's criterion, two light sources can be resolved by an optical system if the distance between them is more than d_{difr}:

2 2D setup

$$d_{difr} = \frac{1.22\lambda}{2N_A} \qquad (2.1)$$

where λ is the light wavelength and N_A is the numerical aperture of the objective. In the interferometry the diffraction limit has the following meaning: if two objects are at distance smaller than the diffraction limit d_{difr}, the fringe pattern contrast of the objects is low and the objects are not resolved by the interferometer. Using Eq. 2.1 the diffraction limit for the objectives utilised in the 2D TIM setup at wavelength 1.3 µm was calculated. The results are presented in Table 2.1 in the last row.

From the comparison of the diffraction limit and the pixel size it can be seen that the setup resolution is limited by the diffraction limit. However, the high number of pixels is not redundant since it is necessary for the fine sampling of the interference pattern.

2.3.3 Time resolution

The time resolution of the setup is determined by the laser pulse duration. During this time the DUT is illuminated and the interferogram is exposed on the camera. The duration of the laser pulse was measured with the oscilloscope and is shown in Fig. 2.9. From the decrease of the intensity to $1/e^2$ the typical duration of the laser pulse was found to be $2\tau = 3.4$ ns for laser 1 (called Vibrant) and $2\tau = 5.3$ ns for laser 2 (called Opolette). The time resolution of the setup is therefore considered to be 5.3 ns.

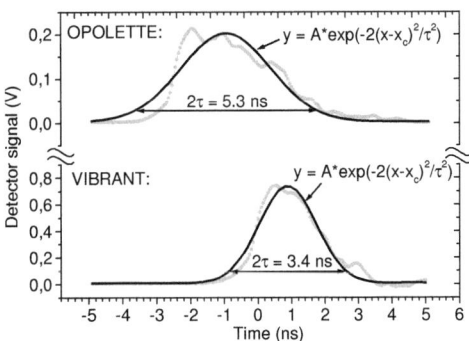

Fig. 2.9. Typical duration of the laser pulse of laser 1 (Vibrant) and laser 2 (Opolette). The profile is obtained by averaging of 50 pulses and it is fitted by a Gauss function.

2.3.4 Laser specifications

The two lasers, Opolette 355 II and Vibrant 355 II manufactured by Opotek, are used in the setup shown in Fig. 2.2b. Both are based on the parametric generation of the light. The advantage of the OPO (optical parametrical oscillator) lasers is (i) the high power (in comparison to a diode laser), which is necessary to record the interferogram in the vidicon camera, (ii) the low coherence length (iii) a good space coherence and (iv) the possibility to tune the wavelength in a wide range, since for some applications and investigations different wavelengths are required. The disadvantage is the pulse to pulse instability and high costs.

The OPO of the laser system consists of a non-linear optical crystal placed within an optical cavity. A high power laser beam (third harmonic of the pulsed Nd:YAG laser – 355 nm) is injected into the cavity to pump the OPO crystal. As a result of the non-linear interaction between the crystal and the laser beam, the signal ν_s and idler ν_i frequencies are generated. Here the frequency condition $\nu_p = \nu_i + \nu_s$ must be fulfilled, where ν_p is the frequency of the pump beam. These frequencies depend on the orientation of the crystal and they are changed by turning the crystal (rotation range is 20 degree). The two beams are orthogonally polarised and can be simply separated by a polarizer. The angle at which the signal and idler wavelengths are equal is called degeneracy point (710 nm) and here the generation efficiency is the lowest. The two lasers differ mainly in the output power and pulse-to-pulse stability. Their specifications are shown in Table 2.2.

Table 2.2. Specifications of the lasers used in the 2D TIM setup.

	Vibrant laser	**Opolette laser**
Tuning range	400-2400 nm	400-2400 nm
Output energy of the idler beam at 1300 nm	4 mJ	0.3 mJ
Linewidth in whole tuning range (specified)	3-5 cm^{-1}	3-5 cm^{-1}
Coherence length (air, measured)	2 mm	2 mm
Pulse duration (measured)	3.4 ns	5.3 ns
Maximal pulse repetition frequency	10 Hz	20 Hz
Pulse-to-pulse stability (standard deviation, measured)	4 %	13 %

The coherence length of the lasers $L_{coh} \cong 2$ mm (in the air) was measured by shifting of the reference mirror and measuring the fringe amplitude. Decrease of the fringe amplitude to $1/e^2$

2   2D setup

was taken as a criterion. This is consistent with the laser specifications (specified is the linewidth $\Delta = 3\text{-}5$ cm^{-1}, from this the coherence length is $L'_{coh} = 1/\Delta = 2\text{-}3.3$ mm).

2.3.5   Camera specifications

Two cameras are used in the setup. These are the infrared vidicon cameras produced by Hamamatsu, type C2400-03. Their spectral sensitivity is 400-1800 nm. The main advantage of the vidicon camera is the sensitivity in the near IR region (in comparison to e.g. CCD camera) and the low costs (in comparison to e.g. focal plane array (FPA) camera). The disadvantage is the lag of the vidicon tube. Several parameters of the camera that affect the recording of the interferograms are discussed bellow.

Gamma characteristic. The relationship between the signal output I_{out} and the incident light I_{in} is called gamma characteristic Γ. This relationship is defined by equation $I_{out} = I_{in}^{\Gamma}$. It is linear for $\Gamma = 1$. Originally gamma of the vidicon camera differs from one, but it can be electronically well corrected in a camera controller. This is demonstrated in Fig. 2.10.

Linearity of the camera is very important for the phase extraction. If the gamma differs from one, the interferogram recorded with the camera does not satisfy Eq. 1.10 since higher harmonics of the fringes appear and the phase shift is difficult to extract.

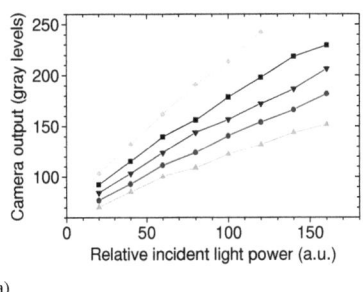
(a)
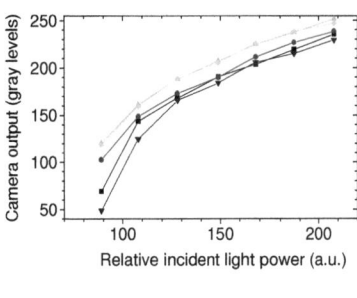
(b)

Fig. 2.10. The gamma characteristic of the camera at five randomly chosen pixels for (a) $\Gamma = 1$ (after correction) and for (b) $\Gamma = 0.3$ (before correction).

Lag. Lag is a phenomenon caused when some of the output camera signal lingers after the incident light was interrupted. This is a typical phenomenon for vidicon cameras and it limits the speed of the measurement. Because there is only rough information about the lag in the camera manual (typical value is specified to 1 second), it was necessary to rate this phenomenon.

The decreasing of the intensity from frame to frame after the laser pulse is shown in Fig. 2.11a. In some cases the first frame can be overexposed and the second or third frame can be only processed. Therefore during the measurement the acquisition of more than one frame per one laser pulse in necessary.

The lag was measured by mechanical interruption of the incident light, see Fig. 2.11b. After the light is interrupted, the camera output decreases from frame to frame. The decrease of the output to 1/e defines the lag.

The camera controller allows controlling of the camera gain, offset and sensitivity. These three parameters define the camera output by relation $I_{out} = (I_{in}*\text{SENSITIVITY} - \text{OFFSET})*\text{GAIN}$. The lag of the camera depends on the sensitivity parameter and this dependence is shown in Fig. 2.11c. Since the typical value of the sensitivity is 8-9 levels during the measurements, the recommended repetition frequency of the experiment is 1 Hz. Increasing of the repetition frequency results into a superposition of the previous damped interferogram with the new one.

Resolution. The video output signal from the camera is coded in CCIR (PAL) system with frame rate 25 frames per second. This signal is led to the IMAQ (image acquisition card) NI PCI-1407 manufactured by National Instruments. Here the analogue signal is digitised to 8 bit grey scale in format of frames 768x576 pixels.

Noise. The standard deviation of the electronic noise of the camera pixels was found to be 2-4 grey levels (depending on the gain) from frame to frame.

Stability. If the camera is exposed under constant light condition, the signal on the camera slowly increases. After 2-3 minutes the signal increases up to 20 grey levels. This property is important to know if some experiments under constant light conditions are done.

2 2D setup

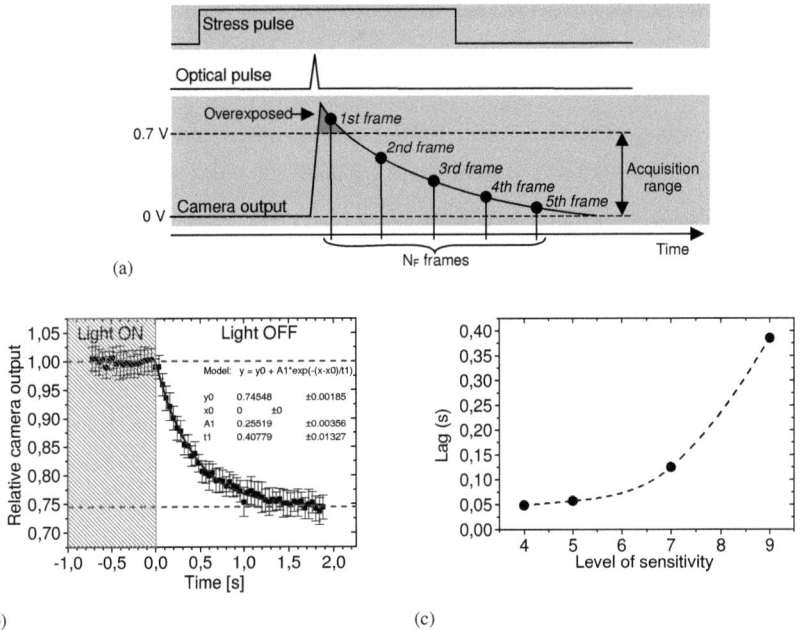

Fig. 2.11 Lag of the vidicon camera. (a) The response of the vidicon camera to a short laser pulse. First several frames can be overexposed, the intensity of the frames (camera output) slowly decreases after the pulse. (b) Measurement of the camera lag by switching off the light. The level of sensitivity was set to 9 in the camera controller. (c) Interpolated dependence of the camera lag on the level of sensitivity. The lag levels correspond to the decrease of the camera output to 1/e.

2.3.6 Electrical stressing

In the setup three pulsers are currently available. The first one is a high voltage (HV) electronic switch SV4000-P produced by DEI (Direct Energy Inc). The voltage pulses generated by this switch are rectangular with risetime approximately 20 ns and maximal amplitude 4 kV. Using 1 kΩ and 50 Ω resistors the voltage pulse is converted to a current pulse, see Fig. 2.1a. A current probe (optionally a voltage probe) monitors the current (voltage) through the device on the oscilloscope. The pulse amplitude through the DUT can rise up to several amperes and the pulse duration can be set from 150 to 2000 ns.

If a shorter risetime (~7 ns) or a bias voltage is required, the Agilent Technologies pulser 8114A is available. This device generates stress pulses up to 100 V or 2 A into 50 Ω. Its load impedance is adjustable.

For the shortest pulses from 5 ns to 100 ns the Avtech pulser AVL-3AH-C is used. The risetime of these pulses is bellow 1 ns and maximal amplitude 450 V into 50 Ω.

2.4 Effect of the DUT on the measurement

The beam reflected from the DUT carries the information about the refractive index changes, but also it is modulated by the DUT reflectivity and surface too. Some of these properties can degrade the interferogram, but they can not be changed without influencing on the DUT functionality. Therefore all the properties, that effect the interferogram, have to be classified and their influence on the interferogram has to be analysed.

2.4.1 Sample backside surface roughness (polishing)

The interferogram quality is strongly influenced by the roughness of the DUT backside surface. The surface imperfections of size larger than the wavelength deflect the probe beam, see Fig. 2.12.

The imperfections or scratches with the size comparable or smaller than wavelength of the laser (1.3 μm) scatter the light and cause the modulation of the fringe pattern contrast. An example is shown in Fig. 2.13. The light scattering causes the spatial modulation of the accuracy of the extracted phase. In darker regions (low contrast fringe pattern) the extracted phase is more noisy (up to 2 times) and contains more artifacts (see Fig. 2.13). For such interferograms an additional smoothing of the extracted phase is necessary. Also the scratches cause spreading of the $a(u, v)$ spectrum component (see Eq. 1.17) in direction perpendicular to the scratches, which makes spectrum filtering more difficult. Best results are obtained if the fringes are oriented perpendicular to the scratches.

2 2D setup

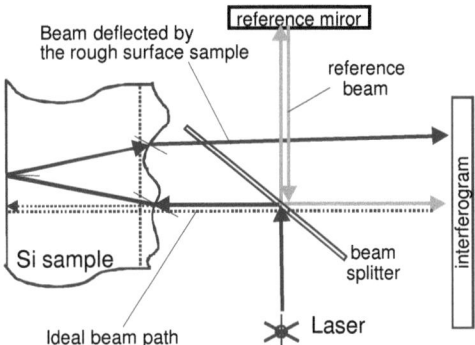

Fig. 2.12. Deflection of the probe beam by rough surface. The surface causes a deformation of the probe beam wave front and results into a deformed interferogram.

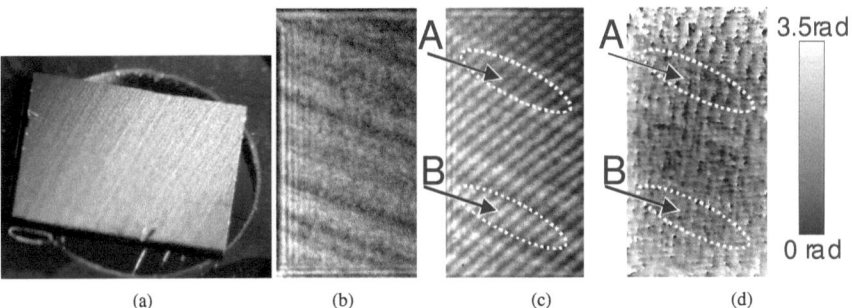

Fig. 2.13. Scratches of the sample backside. (a) Scratches after the mechanical polishing (sample size 3x4mm). The scratches that are visible in (a) modulate the intensity of the reflected probe beam (b) and fringe pattern contrast (c). Due to this the area 'A' has up to two times lower fringe amplitude than the area 'B'. Figure (d) shows the extracted phase map. The phase map in 'A' region has approx. two times higher noise than in region 'B'.

In order to avoid the problems with rough surface the roughness of the sample surfaces should be below the diffraction limit. The polishing with diamond paste (1 µm grain diameter) is sufficient. The chemical-mechanical polishing is also suitable.

2.4.2 Reflectivity of the sample

The DUT basically consists of silicon substrate, SiO_2 layer and metal layers. An electrical contact between two metal layers is introduced by metal contacts called "via". An electrical contact between the metal layer and doped Si region is introduced by metal contacts called "contacts". Layout and structure of these and any other layers (e.g. passivation layer, field oxide, polysilicon etc.) depend on the device technology. The electrical signal is introduced from outside to the metal layers by the contact pads.

The high reflectivity of the layers is important to achieve satisfactory fringe amplitude and consequently the phase accuracy. The reflectivity depends on the layers geometry, thickness and complex refractive index. An example of the interference image and sample cross section is shown in Fig. 2.14. Accurate measurements of the reflectivity of the different layers for SPT5[1] technology were performed on shallow junction SPT5 DEMAND[2] test devices with a focused laser beam, see Fig. 2.15. These devices differ in the layout above the active area only. The results are summarised in Table 2.3.

The best reflectivity exhibits the power metal, metal1 and metal2. Interferogram has the highest S/N ratio (~ 10) in these regions and thus the maximal precision of the extracted phase (up to 0.1 rad). The areas with contacts exhibit the worst reflectivity. The contacts consist of metal rectangles having size 0.6x0.6 μm^2 and 0.8 μm distance in between (bottom view). They scatter the light and therefore are observed as dark regions in the interferogram. In these regions the phase precision is the lowest (around 0.8 rad). The diluted contacts exhibit a better reflectivity and the phase errors are smaller than for the normal contacts (~ 0.3 rad). Scattering and deflection of the focused light on the metal edges or layers with size smaller than the beam spot cause artificial peaks that are higher or lower than the real reflectivity of the layer. It has also been found that the reflectivity of the sample is low at the places with deposited polysilicon layer (material for the gate electrode).

---

[1] SPT5 – Smart Power Technology, generation 5

[2] DEMAND – An Integrated Design Methodology for Enhanced Device Robustness

2 2D setup

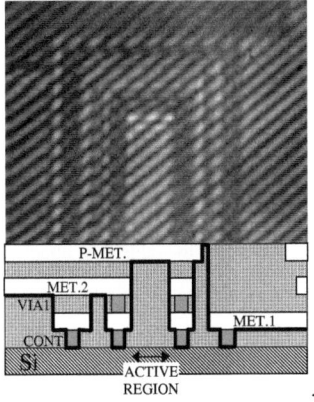

Table 2.3. DUT reflectivity.

Layer	Relative reflectivity (± 5 %)
Power metal	1
Metal 1	0.74
Metal 2	0.61
Contacts	0.09
Contacts diluted	0.32
SiO_2	0.23

Fig. 2.14. The interferogram of the test sample SJ1 (SPT5) and its cross section. The thick black line copies the surfaces that reflect the most of the laser beam intensity.

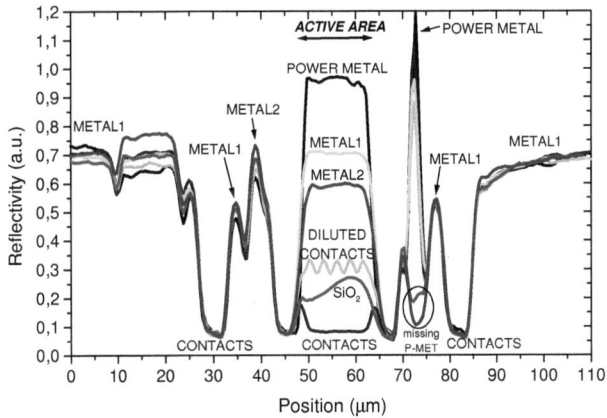

Fig. 2.15. The reflectivity measurement of the SJ1-SJ9 test devices (SPT5). Measurement was done with a focused laser beam of wavelength 1.3 µm in the heterodyne setup. The devices differ in the layout above the active area. For diluted contacts the contact structure is seen.

2 2D setup

2.4.3 Lateral geometry of the device and fringe discontinuity

At different regions of the DUT the laser beam is reflected from the different layers: metal1, metal2, power metal, etc. Therefore at the edges of the two layers the fringes exhibit distortions or discontinuities due to the abrupt changes of the phase. An example of such fringe discontinuities is shown in area A in Fig. 2.16a.

Another example is an area B (see Fig. 2.16a), where the fringes at the structure edge E look like continuous. Nevertheless there is an abrupt phase jump at the edge E of about 28 rad, as calculated from the distance between the metal1 and metal2 layers. To show how the fringes would look in the sample with a finite phase gradient at the edges, but the same amplitude, a fringe simulation was performed, see Fig. 2.16b. The real path of the fringe (dashed line) can be traced because of the gradual phase changes at the edge.

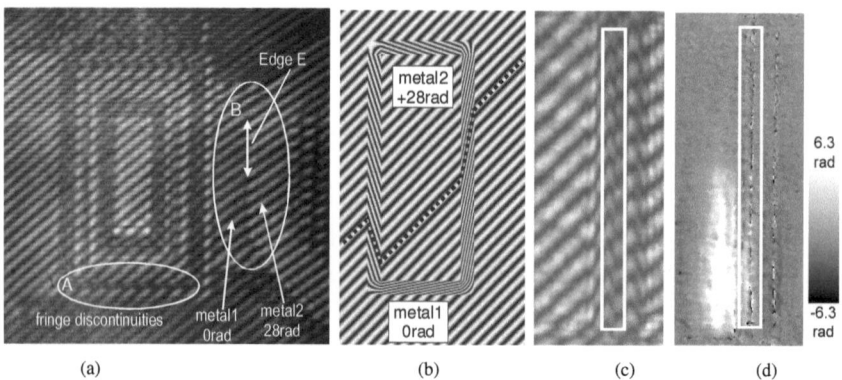

Fig. 2.16. Fringe discontinuities at layout edges. Ovals A and B mark two experimental examples of the fringe discontinuities at the layout structure in interferogram (a). In area A, a fringe kink is clearly seen. In area B the fringes are apparently continuous despite a 28 rad phase jump. Interferogram (b) shows the simulation of the area B for the case of gradual phase changes at the edges. Interferogram (c) is a detail of the edge, which in the phase results into localised phase distortions (marked by a rectangles).

In the experimental interferogram the structure edges have the sizes less than the diffraction limit. Also the laser beam is scattered on the structure edges which decreases the S/N ratio in these areas. These factors cause the ambiguity of the fringe continuity at the edges.

2 2D setup

During the interferogram processing the phase extraction algorithm will connect the closest fringes of metal1 and metal2 and calculate the phase jump as a jump with respect to modulo 2π.

It should be noticed that the useful phase signal is the phase difference between the phase profile with and without stress. Therefore the above type of discontinuities have a marginal influence on the phase extraction accuracy. Finally only the localised phase errors along the edges increase the noise, see Figs. 2.16c, d.

2.4.4 Cellular device structure and the phase extraction accuracy

By cellular structure we mean a structure with repeating small geometrical features with a size comparable to the fringe thickness or to the diffraction limit.

DMOS device: An infrared image of a multi-cell DMOS device with a cell size 5x5 μm^2 and 5 μm distance between the cells is shown in Fig. 2.17a. The space between the cells and the cell size (5 μm) are larger than the diffraction limit (1.9 μm). The edges of the metal layers in a cell scatter the light and distort the fringe pattern (see Fig. 2.17b). The extracted heated region can be resolved with a good accuracy, see Fig. 2.17c. A small artifact occurs at a place with a low reflectivity.

Bonding pads: A bonding pad has a cell size of 4x4 μm^2 (similar to DMOS cells) and a small space in between (1 μm). The grey rectangles in the Fig. 2.17d are the holes in metal1. The laser light penetrates through these holes deeper into the SiO_2 layer and the metal2 layer reflects it. The interferogram of such cell structure is shown in Fig. 2.17e. The inclined fringes are continuous with (periodical) distortions by the scattering at the cells edges. The cross section of the extracted phase shift is shown in Fig. 2.17f. The space between the cells is less than the diffraction limit and therefore the rectangular contour of the cell is not resolved in the interferogram. The periodicity of the cell structure can however be resolved. Nevertheless, some phase peaks exhibit one and some two local maximas. It is expected that if the active region was located in this area, the subtracted phase shift accuracy would be very bad (worse than 1 rad), since the one peak / two peak feature may be very sensitive to fringe position.

2 2D setup

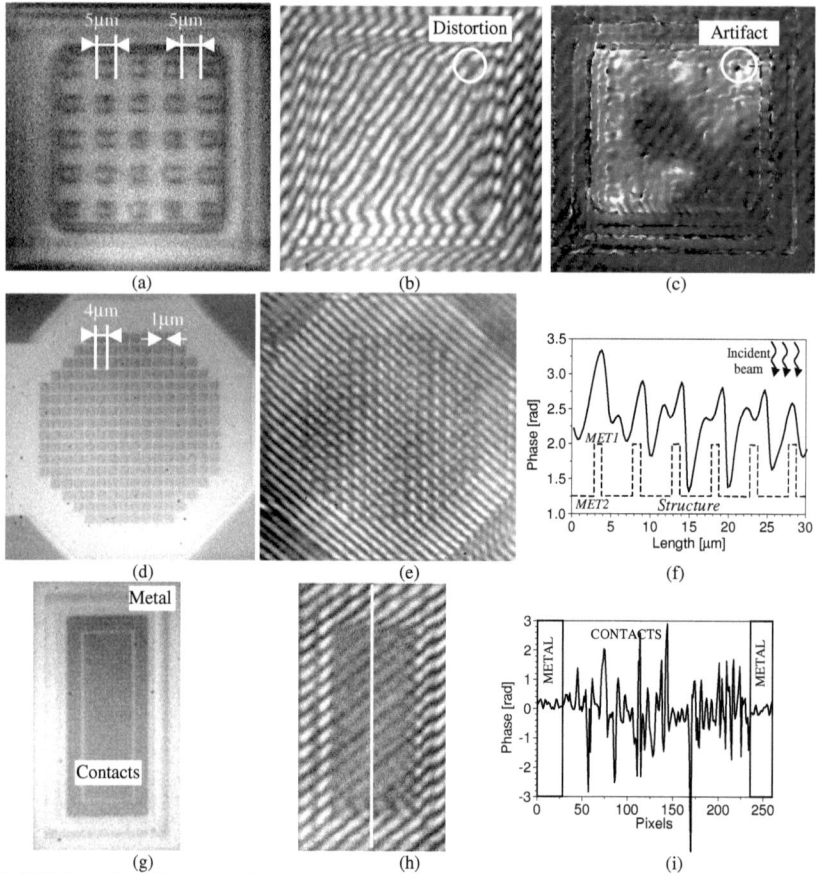

Fig. 2.17. Broadband light image of (a) DMOS cell structure, (d) bonding pad structure and (g) contacts in active region. The images were made using an immersion oil objective (NA = 1.3) for better resolution of the cell structure. The corresponding interferograms are (b), (e) and (h), respectively. In the interferogram (b), the bending and amplitude modulation of the fringes is visible. In (e) the structure disturbs the fringe continuity and straightness. In interferogram (h) the fringes on contacts are parallel but have very low amplitude in comparison with the fringes on metal. Figure (c) shows the extracted phase shift for DMOS, (f) shows the extracted phase of the bonding pad cells and the theoretical phase profile and (i) shows the extracted phase along the white line in (h).

Contact plugs: The contact plugs, that connect Si and metal1, have dimensions (size $0.6 \times 0.6 \ \mu m^2$ and 0.8 μm space in between) below the diffraction limit (see Fig. 2.17g-i). The contact structure is not resolved in the picture, therefore the contact area looks like a dark rectangle (see Fig. 2.17g). The contacts do not distort the fringes, but the fringe pattern contrast in this 'sub-diffraction' area is approx. 5 times lower than in the metal region (see Table 2.3). The latter effect decreases the S/N ratio of the extracted phase. An example of the extracted phase cross section is given in Fig. 2.17i. The phase noise is much higher in the contact area than at the metal (approx. 4 times).

2.5 Phase extraction

After the interferogram is recorded, the phase shift has to be extracted from the interferograms. As it was shown in Chapter 2.4, the fringe amplitude and shape is modulated by the DUT structure, reflectivity and backside polishing. From this reason, some phase extraction algorithms can not be used in our case (see Chapter 1.4.8), for example: (i) *fringe skeletonizing* method is very sensitive to the fringe amplitude modulation and fringe discontinuities; (ii) *phase sampling* method requires recording of at least three phase-shifted interferograms, which may not differ in the intensity distribution (requires stabile pulse source and digital camera) and additionally it is not suitable for investigation of unrepeatable phenomena. We have established, that the most rigid method against the fringe modulation by DUT and even against the pulse-to-pulse variations of the laser beam is the method based on the 2D *Fourier transform*. This method has been therefore implemented, analysed in details and adapted to our particular case in order to automate the phase extraction and to get result with the best S/N ratio.

There is some software on the market, which uses the 2D FFT method for the phase extraction. This is e.g. IDEA (Interferometrical Data Evaluation Algorithms) developed in TU Graz, Austria (http://optics.tu-graz.ac.at) or Fringe Processor developed in Bremer Institut fuer angewandte Strahltechnik, Germany (www.bias.de, www.fringeprocessor.com). Both support also other phase evaluation techniques. The software includes image-processing procedures for improvement, segmentation, analysis, basic mathematical operations and high-end visualisation of images. The only disadvantage of such very sophisticated software is that the phase evaluation can not be calculated automatically. Automatic phase extraction rapidly increases the speed of the phase extraction and thus the whole investigation process of the DUT. Therefore a home-made software was programmed. Further advantage of a home-made software is that it is specially designed for our particular case and therefore it is more rigid against the artifacts typical for our interferograms.

The phase extraction process can be influenced in two points: during the spectrum filtering and during the phase unwrapping. Therefore the effort was concentrated on these two processes.

In the following part an optimal phase extraction sequence for evaluation of interferograms recorded in our 2D TIM setup is presented. We added several novel features into the standard phase extraction sequence (FFT – spectrum filtering – inverse FFT – unwrapping), which simplify the phase extraction, increase the speed of phase extraction and limit the number of artifacts in the final phase image. Next the details about the spectrum filtering like noise filtering, overlapping of spectral components, effect of fringe thickness on the phase extraction and optimal spectrum filter will be described. Then the phase unwrapping and some other factors that have an effect on the phase measurement will be discussed.

2.5.1 Phase extraction sequence

The flow chart of the optimised phase extraction sequence is shown in Fig. 2.18. The novelties in comparison with standard phase extraction technique are included in Steps 3, 6 and 7. The fully automated phase extraction starts with Step 2 and ends with Step 7. It is marked by a grey arrow in Fig. 2.18.

STEP 1: In the beginning the interferograms that will be processed are selected. Several interferograms of the stressed device are chosen and one reference interferogram, which was recorded before the stress pulse application.
STEP 2: The 2D spectrum of the interferograms is calculated.
STEP 3: The filtering of the spectrum components $a(u, v)$ and $c^*(u, v)$ is done (see Eq. 1.17). The filtering and its consequences are investigated in detail in order to design a novel optimal spectrum filter, as will be shown in Chapter 2.5.2.
STEP 4: The inverse FFT of the filtered spectrum is calculated.
STEP 5: Phase mod 2π is obtained from the inverse FFT (see Eq. 1.18).

STEP 6: Subtraction of the phase mod 2π corresponding to the stressed and reference interferogram is performed. Phase mod 2π corresponding to the stressed interferogram contains the information about the refractive index changes, device vertical topology and inclination of the reference wave (equivalent to fringe carrier frequency). By subtraction of the phase mod 2π corresponding to the reference interferogram the device topology and inclination of the reference wave is removed. Thus the resulting phase contain information just about the refractive index changes.

STEP 7: Before the unwrapping starts, the phase is pre-processed. Thanks to the pre-processing the unwrapping of the whole phase is not necessary and this rapidly decreases the probability of the error creation and increases the speed of the unwrapping. Details about the pre-processing are explained in Chapter 2.5.3.1.

STEP 8: Unwrapping of the user-selected areas is the last step of the phase extraction.

STEP 9: Before the result is saved, it can be post-processed in order to remove the noise, offset etc.

In our software we implemented the automated phase extraction procedure described above, but also a manual mode. In the manual mode the user has a full control over the process at any step of the phase extraction. This is helpful in the case of low quality interferograms, like it is for example in the case of DMOS (see Chapter 2.8.3). Typical duration of the automatic phase extraction from one reference interferogram and one interferogram (size 768x576 pixels) of the stressed device in Pentium IV, 1.8 GHz, 512 MB RAM is 10 seconds. For analysis of any further interferogram of the stressed device this duration increases by about 6 seconds per interferogram. Details on the phase extraction software are in Appendix.

2 2D setup

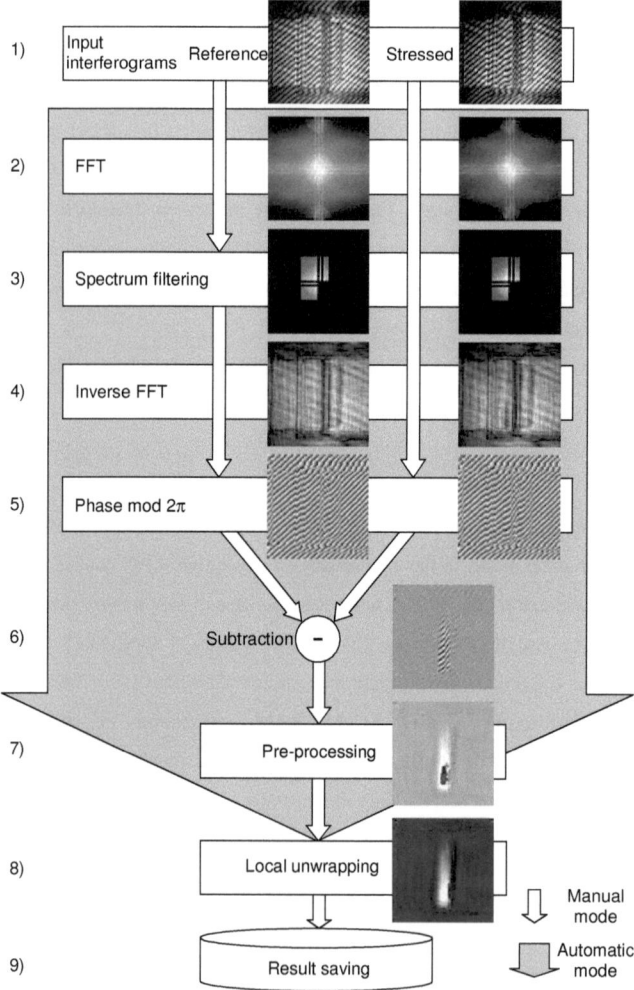

Fig. 2.18. Flow chart of the optimised phase extraction using the 2D FFT method. The automatic phase extraction is marked by grey arrow (automatic mode), the single steps of the manual mode are marked by white arrows.

2.5.2 Spectrum filtering

The spectrum components $a(u, v)$ and $c^*(u, v)$ (see Eq. 1.17) [Kre96] have to be filtered out to calculate the phase. Since these functions usually overlap with $c(u, v)$ in spite of using the fringe carrier frequency, their filtering is not trivial. A spectrum filter usually blocks part of the $c(u, v)$ component and transmits part of the $a(u, v)$ and $c^*(u, v)$ components. This results into undulations in the phase. To suppress the undulations the spectrum filtering was analysed and an optimal filter was designed [Dub04a].

The interferogram is a discrete array of finite number of pixels. The discreteness of the interferogram pixels leads to the discreteness of the spectrum. The spectrum bandwidth is limited to frequencies smaller than the Nyquist frequency $f_c = 1/2\Delta$ [Otn78], where Δ is the sampling interval of the interferogram, i.e. Δ^{-1} is the sampling rate. In our case $\Delta = 1$ pixel. The finite number of the pixels leads to the finite number of the spectrum points. The spectrum is defined for frequencies $f_n = n/N\Delta$, where $n = -N/2, \ldots N/2$ and N is the number of the pixels in interferogram [Otn78]. Similarly in 2D space the spectrum is defined in frequencies (f_{nx}, f_{ny}), where n_x and n_y vary in range $-N/2$ to $N/2$. The fringe carrier frequency will be marked like (f_{Fx}, f_{Fy}). The number of the pixels of the 2D spectrum is the same as the number of the pixels of the interferogram. In our case it is important to understand the relation between the fringe period $T_{Fi} = 1/f_{Fi}$ (i = x, y) defined in pixels of the interferogram and the position in the calculated spectrum (n_{Fx}, n_{Fy}) in pixels:

$$n_{Fi} = Nf_{Fi}\Delta = \frac{N}{T_{Fi}}, \text{ where } i = x, y. \quad (2.2)$$

The finite size of the interferogram is equivalent to convoluting the Fourier transform with $\sin(i)/i$ function [Otn78]. This results in "smearing" of the spectrum. For example, the parallel fringes is Fig. 2.19a are transformed into two "stars" $c(u, v)$ and $c^*(u, v)$ centered around the fringe carrier frequency, see Fig. 2.19b. From this it is clear, that the spectrum components $a(u, v)$, $c(u, v)$ and $c^*(u, v)$ have the best separation if the fringes are oriented diagonally.

In following chapters the noise filtering, overlapping of spectral components, effect of fringe thickness on the filtering of the spectrum components $a(u, v)$ and $c^*(u, v)$ and the optimal spectrum filter will be studied.

2 2D setup

2.5.2.1 Noise filtering

The noise in the phase mod 2π disturbs the phase unwrapping, therefore it has to be decreased before the unwrapping process starts. There are three ways how to decrease the phase noise and to increase the phase sensitivity: the filtering of the high frequencies of the spectrum, the averaging of several images and the smoothing.

<u>High frequency filtering.</u>

A simulated noiseless interferogram with Gaussian phase object is shown in Fig. 2.19a. In this case overlapping of $a(u, v)$, $c(u, v)$ and $c^*(u, v)$ can be neglected (see Fig. 2.19b) and the noise effect can be investigated here. Two spectral filters depicted in Figs. 2.19c, d were applied to the spectrum. Filter 1 (Fig. 2.19c) filters out the frequencies with the highest amplitude belonging to $a(u, v)$ and $c^*(u, v)$. Filter 2 (Fig. 2.19d) filters out everything except the frequencies with the highest amplitude belonging to $c(u, v)$. Using of both filters leads to phase without noise, see cross section in Fig. 2.19f.

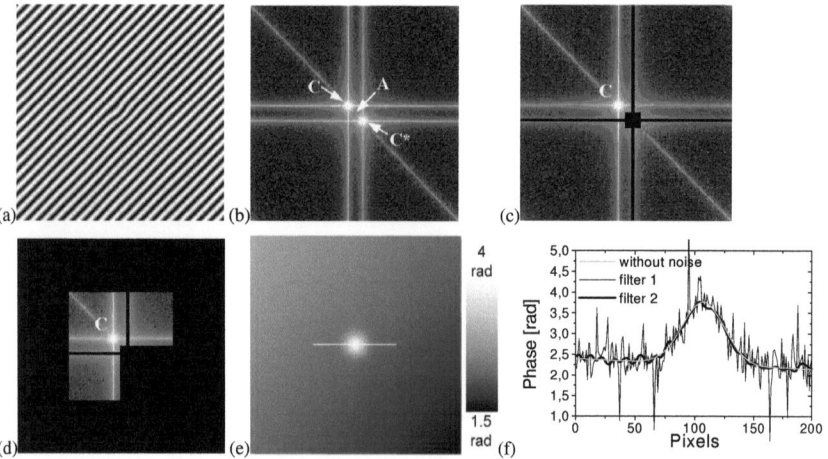

Fig. 2.19. (a) Simulated interferogram of a Gauss heat source, (b) its spectrum, (c) spectrum filter 1, (d) spectrum filter 2, (e) extracted phase, (f) cross section of the phase along the white line marked in (e). The cross section for the interferogram without noise is plotted in grey colour. The cross section for the interferogram with S/N = 4 and for the filter 1 is plotted with a thin line and for the filter 2 with a thick line.

2 2D setup

If a noise is added to the interferogram in Fig. 2.19a, the application of the filter 1 leads to a noise in the phase, see Fig. 2.19f. Filter 2 removes the high frequency noise components from the spectrum, but the low frequency noise components lead to the low frequency phase undulations, see Fig. 2.19f. The undulations have lower amplitude than the noise resulting from filter 1 and therefore do not disturb the unwrapping.

Filtering of high frequencies limits the space resolution according to the cut-off frequency. Assuming the Eq. 2.2, if the cut-off frequency is in half of the spectrum ($n_i = N/4$, see Fig. 2.19d), the spatial resolution of the phase decreases from one pixel to two pixels (T_{Fi} is changed from 2 to 4). Often the size of one pixel is below the diffraction limit (see Table 2.1, size of a pixel for maximal zoom for 50x objective is 0.3 µm and diffraction limit is 1.9 µm), therefore the filtering does not effect the setup space resolution.

Smoothing.

Smoothing of the interferogram is done by applying median, average or Wiener filter. The noise in the phase thus decreases. But if during the spectrum filtering the high frequencies are filtered out (see filter 2 in Fig. 2.19d), the effect of smoothing is negligible.

If the interferogram contains the noise, the noise is present in the spectrum too. However, smoothing of the spectrum is not allowed, since it changes the amplitude of the phase shift, see Fig. 2.20a.

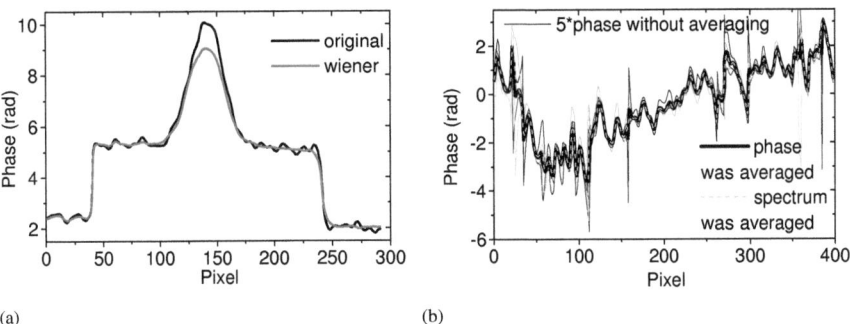

Fig. 2.20. Smoothing of the spectrum. (a) Comparison of the extracted phase if the spectrum is smoothed by the Wiener filter. Due to the Wiener filter the amplitude of the phase peak is lower. (b) Phase cross section for five interferograms and the effect of averaging. Averaged were the phases and their spectrums. Averaging of the phases and spectrums leads to an identical result.

2 2D setup

Averaging of several phases

By averaging of the extracted phase of several interferograms the noise can be decreased according to the statistical laws, see example in Fig. 2.20b. However the averaging is not possible when the investigated phase object exhibits unstable behaviour (e.g. it randomly changes the location or amplitude from pulse to pulse). The averaging is a time consuming procedure too.

Noise sources

The electronics of the camera and the video card have a noise. This noise is visible in Fig. 2.41b as random fluctuations superposed on the signal. These fluctuations are transformed into the phase fluctuations during the phase, see Fig. 2.20b. The level of the phase fluctuations can be estimated by the following equation:

$$\delta\varphi \approx arctg\left(\frac{\delta b}{B}\right) = arctg\left(\frac{1}{S/N}\right) \qquad (2.3)$$

where the $\delta\varphi$ and δb are the standard deviations of the phase and interferogram noises, respectively, B is the amplitude of the fringes, S/N is the signal to noise ratio. Typical experimentally measured values of the S/N ratio and the standard deviations $\delta\varphi$ of the extracted phase in different areas of the SPT5 technology devices are presented in Table 2.4. The areas covered by the metal layers have the highest S/N values (see columns 'Power metal' and 'Metal 1' in Table 2.4) due to high reflectivity. The light scattering on the contact plugs causes the small S/N value (see column 'Contacts' in Table 2.4).

Table 2.4. Noise in the interferogram and phase at different regions on the DUT.

Area	Power metal	Contacts	Metal 1	SiO$_2$
S/N	10	1	5	2.4
$\delta\varphi$ (rad)	0.1	0.8	0.2	0.4

2.5.2.2 Spectrum overlapping

Object with different phase or reflectivity

If objects with phase step or reflectivity modulation are present in the interferogram, the spectrum components $a(u, v)$, $c(u, v)$ and $c^*(u, v)$ spread and start to overlap. This makes problems by filtering of $a(u, v)$ and $c^*(u, v)$ components and cause undulations of the phase. To design a spectrum filter, the effect of the device on the spectrum was investigated.

A simulated interferogram representing a rectangular step region is shown in Fig. 2.21a and its spectrum in Fig. 2.21b. The overlapping of the spectral components is indicated. For any chosen filter either a part of $a(u, v)$ and $c^*(u, v)$ components remains in the spectrum, or a part of $c(u, v)$ component is filtered out. In both cases, undulations in the extracted phase are created. The cross section of the extracted phase in Fig. 2.21c along the marked line is shown in Fig. 2.21d. The undulations are highest near the object edges. The same holds for the objects of different reflectivity, see Fig. 2.22. It was also found that the more objects are in the interferogram, the higher are the undulations.

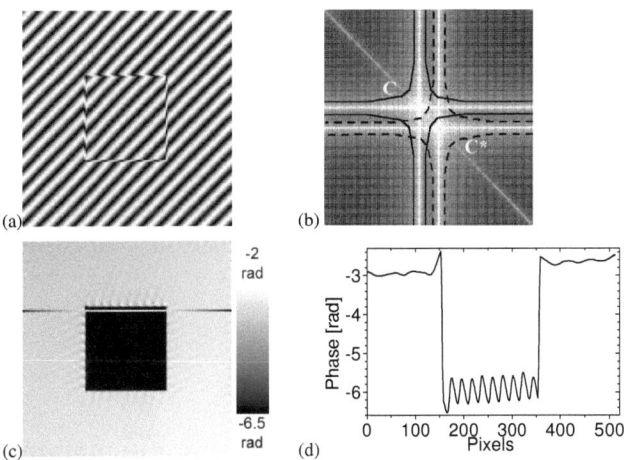

Fig. 2.21. (a) Simulated interferogram of a rectangular structure, (b) its spectrum, (c) extracted phase and (d) cross section along the line marked in (c).

2 2D setup

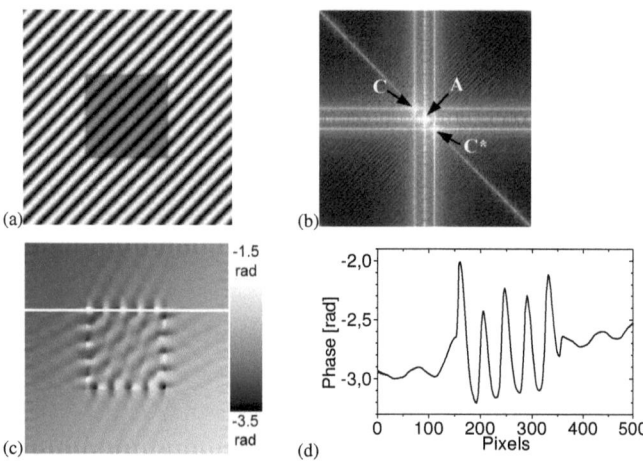

Fig. 2.22. (a) Simulated interferogram of a dark rectangle, (b) its spectrum, (c) extracted phase and (d) cross section along the line marked in (c).

Since the undulations are related to the device structure, they are similar in the reference interferogram and in the interferogram of the stressed device. If a laser source with high pulse-to-pulse stability was used, the undulations are subtracted in the region of constant refractive index, see Fig. 2.23. Only in the region where the refractive index was changed during the stress pulse, the undulations are not fully subtracted.

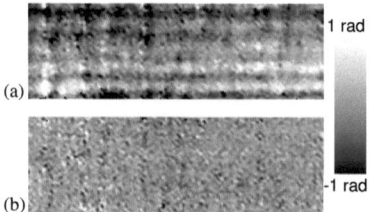

Fig. 2.23. (a) The phase with undulations. Phase was calculated from a single reference interferogram (experimentally measured). (b) The phase without undulations, only the random noise is present. Phase was obtained by subtraction of two reference phases.

2 2D setup

Fringe thickness

The overlapping of the spectrum components $a(u, v)$, $c(u, v)$ and $c^*(u, v)$ is a function of the fringe thickness. Therefore the effect of the fringe thickness on the undulations was investigated.

The Gauss distribution of the phase with amplitude of $0.5\,\pi$ and thickness of 40 pixels was used for the simulation of the phase object (e.g. the heating area). The fringe pattern with diagonal fringes and 10 pixels in period was chosen. The interferograms, spectra and the extracted phase for the case of thin and thick fringes are shown in Figs. 2.24 and 2.25, respectively.

In the case of thin fringes, the spectrum components $c(u, v)$ and $c^*(u, v)$ are well separated from each other (see Figs. 2.24b, c) and the extracted phase shift matches well to the original Gaussian profile (see Figs. 2.24d). In the cross section in Fig. 2.26 the extracted and the simulated phase are coinciding.

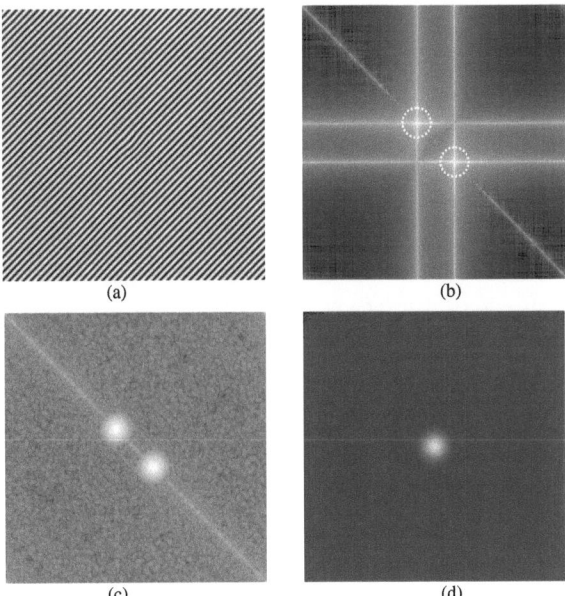

Fig. 2.24. (a) Simulated interferogram with thin fringes. The size of the interferogram is 512x512 pixels. The period of the fringe pattern is 10 pixels. (b) Interferogram spectrum with marked spectral components corresponding to the simulated phase object. (c) Difference of the spectrum of the interferogram with and without phase object, the spectral components related to the phase object are well seen. (d) Extracted phase.

2 2D setup

In case of the thick fringes (see Fig. 2.25) the signal $c(u, v)$ and the conjugated $c^*(u, v)$ spectrum components overlap (Fig. 2.25b, c) and can not be distinguished, which leads to a deformation of the extracted phase and phase undulations, see Fig. 2.25d. This reduces the space and phase resolution. The distortions are well seen in the cross section in Fig. 2.26.

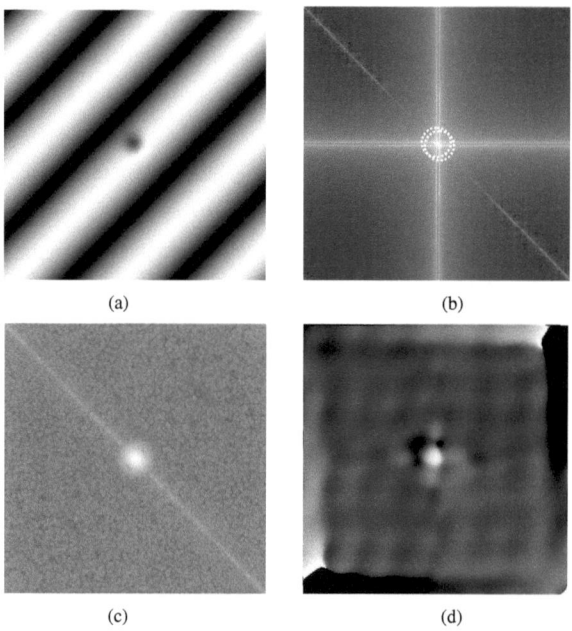

Fig. 2.25. (a) Simulated interferogram with thick fringes, (b) interferogram spectrum with marked spectral components corresponding to the simulated phase object, (c) spectral components related to the phase object and (d) extracted phase.

The effect of different fringe thickness on the extracted phase in the case of a measured interferogram is shown in Fig. 2.27. In the case of thin fringes (Fig. 2.27a) the rectangular heated region is well resolved with sharp edges and low undulations (see Fig. 2.27b). In the case of thick fringes (Fig. 2.27c) the edges of the heated region are less sharp and the phase exhibits more low frequency undulations, which reduces the phase sensitivity.

2 2D setup

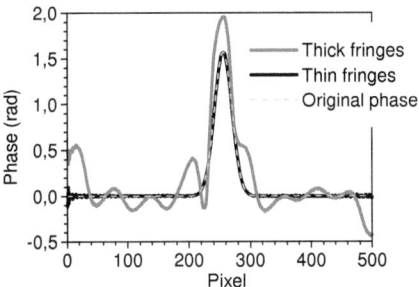

Fig. 2.26. Cross section of the simulated Gaussian phase (dashed curve) and the extracted phase for thick and thin fringe interferograms. Big distortions in the case of the thick fringes are observed.

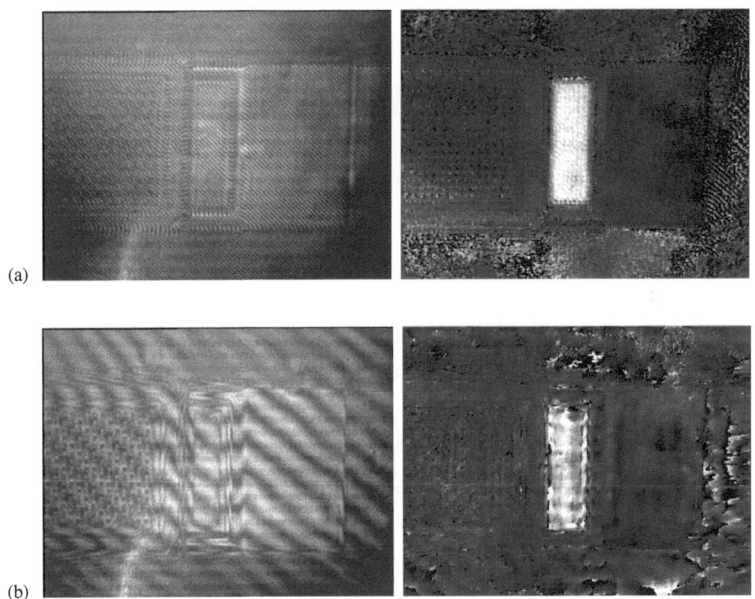

Fig. 2.27. Real interferogram of a rectangular heat source with (a) thin (8 pixels period) and (c) thick (40 pixels period) fringes and the corresponding extracted phase.

2 2D setup

Optimal fringe thickness

The gradient of the phase object is important for the phase extraction. The interferograms with the steep phase objects have a broad spectrum of the phase components $a(u, v)$, $c(u, v)$ and $c^*(u, v)$. Due to this broadening the conjugated components $c(u, v)$ and $c^*(u, v)$ of the spectrum overlap and this causes the artifacts in the phase. In this chapter the criterion for the optimal fringe thickness is established.

An example of the simulated pattern is shown in Fig. 2.28. The simulated phase object has nearly 20x higher amplitude than in Fig. 2.24 but the Gaussian width remains the same. The spectrum of the steep object is wider (compare Fig. 2.24b and Fig. 2.28b) because the spectrum contains components with space frequencies comparable to the gradient of the phase object.

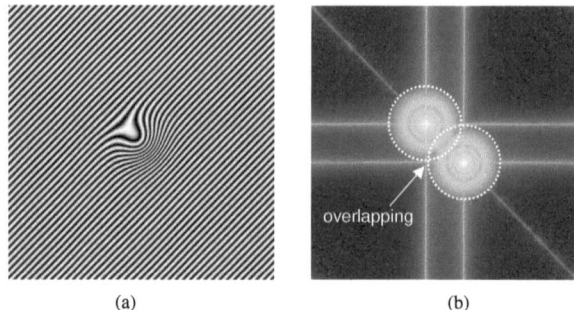

(a)   (b)

Fig. 2.28. Thin fringes and steep phase object. The size of the interferogram (a) is 512x512 pixels, the period of the fringe pattern is 10 pixels. Phase object parameters: height 28 rad, Gaussian distribution with 40 pixels width. (b) Spectrum of the interferogram with marked components of spectrum, which correspond to the simulated phase object.

The criterion for the choice of the fringe period can be estimated. The overlapping of the conjugated spectral components is minimal if the inclination of the reference plane (or "steepness of the fringes", defined by wavenumber $2\pi f_{Fi}$) is higher than the gradient of the phase:

$$2\pi f_{Fi} > \max(|grad\varphi|), \qquad i = x, y \qquad (2.4)$$

where term $\max(|grad\varphi|)$ is a maximum of the phase gradient. The minimum number of pixels per fringe was experimentally found to be 5, which assures a minimum phase sensitivity of 0.5 rad. From that the following criterion for the optimal fringe period is expressed:

2 2D setup

$$5 < T_{Fi} < \frac{2\pi}{\max(|grad\varphi|)}, \qquad i = x, y \qquad (2.5)$$

The phase gradient $grad\varphi$ is in first approximation proportional to the ratio of the phase maximum and the size of the heating source. In practice it means that in the case of the high phase shift or small heating source the thin fringes have to be used. An example of the experimental interferogram with the small heating source (in comparison with the fringe width) is shown in Fig. 2.29. The heating source with the horizontal size of just 8 pixels (i.e. 25 µm) is well resolved.

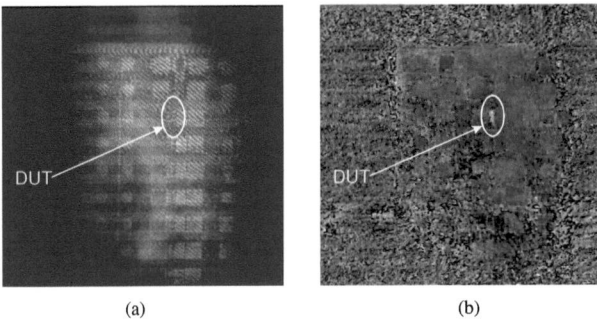

(a) (b)

Fig. 2.29. (a) Example of the thermal mapping with small device size. DUT is marked by oval. Mitutoyo 10x objective was used. Device heated area is visible in (b). The heated area is 25x100 µm (8x22 pixels).

2.5.2.3 Optimal filter design

In previous chapters it was shown, that the undulations in the phase originate from the noise filtering and from the overlapping of the spectrum components (which disable filtering of the whole $a(u, v)$ and $c^*(u, v)$ components). The undulations therefore can not be avoided. They can be only minimised by using a spectrum filter, which is a compromise between the filtering of $a(u, v)$ and $c^*(u, v)$ components and the retaining of the $c(u, v)$ component. Such a filter will be called optimal filter. An optimal filter minimises also the number of the phase defects and the noise in the extracted phase. From this it is clear, that the optimal filter must be adaptive to the particular spectrum and therefore varies for different interferograms.

2 2D setup

The spectrum filter, that we propose, will be explained on the spectrum of the interferogram in Fig. 2.30a. The filter is shown in Fig. 2.30b. It is designed in the following way: the frequencies around the zero ($n_i = 0$) are filtered out up to the one third of the fringe carrier frequency (i.e. $n_{i,\text{down}} = n_{Fi}/3$ for $i = x, y$) (see Fig. 2.30c), that was found to be an optimal value; the cross of the $a(u, v)$ component, the cross of the $c^*(u, v)$ and the whole quadrant of the $c^*(u, v)$ are filtered out (see Fig. 2.30d); the frequencies higher than the upper cut-off frequencies ($n_{i,\text{up}}$) are filtered out (see Fig. 2.30e). The upper cut-off frequencies of the filter are designed in following way: i) the cross of the $c(u, v)$ is smoothed and the position n_i' is calculated, where the signal decreases to 1/3, ii) n_i'' is defined like n_i'' = (n_i' - n_{Fi})*3, iii) if n_i'' > $N/4$, then n_i'' is set to $N/4$, iv) $n_{i,\text{up}}$ is defined like $n_{Fi} \pm n_i$''. At the end, the fields with the pixel value smaller than the mean of the whole spectrum (without filter) are found and the edges of these fields are smoothed. These fields carry more noise than the signal, so to suppress the noise even more they are filtered out too (filtering out means to set the value to the zero), see Fig. 2.30b.

The proposed filter filters well the $a(u, v)$ and $c^*(u, v)$ components if (i) the fringes have diagonal orientation (then the spectrum components are best separated), (ii) the fringe thickness is smaller than the size of the objects (see Fig. 2.29 and Chapter 2.5.2.2) and (iii) the tilt of the reference mirror is higher than the gradient of the phase object (see Eq. 2.4). Then the spreading and overlapping of the spectrum components is minimal.

The spectrum filtering becomes more difficult if the interferogram is over- or underexposed. This causes higher harmonics in the spectrum, see $2n_{Fx}$, $2n_{Fy}$, $3n_{Fx}$, $3n_{Fy}$ in Fig. 2.31. If the higher harmonics are not filtered out, artificial fringes appear in the phase. Filtering out of the higher harmonics is not included in the proposed spectrum filter. To avoid the fringes in the phase, additional manual filtering out of these higher harmonics is necessary.

The high frequency spectrum components, which are marked by ellipses on the top and bottom of the spectrum in Fig. 2.31, originate from the camera interlacing. The signal of the vidicon tube is processed in PAL system, which means that the odd lines are read out first and the even lines are read out afterwards. Since the signal in the vidicon tube decreases with time (see Fig. 2.11), the even lines exhibit lower intensity. These high frequencies are filtered out automatically by the proposed spectrum filter, therefore they do not effect the phase shift.

In the phase extraction sequence (see Fig. 2.18) the same spectrum filter is used for any couple of reference interferogram and stressed interferogram. This minimises the phase undulations to minimum, as was shown in Fig. 2.23.

2 2D setup

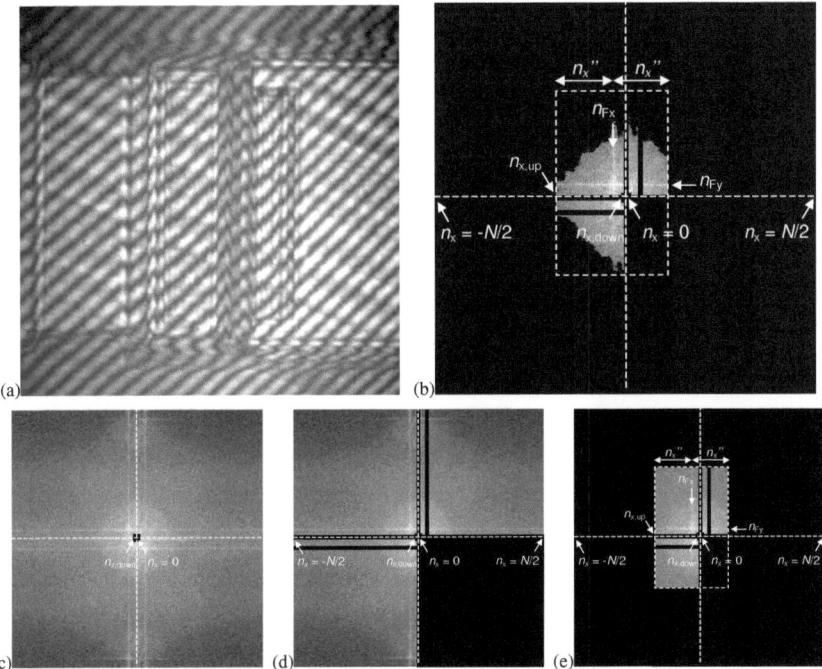

Fig. 2.30. (a) An interferogram. (b) Proposed spectrum filter demonstrated on the interferogram in (a). (c-e) The building of the proposed filter in (b).

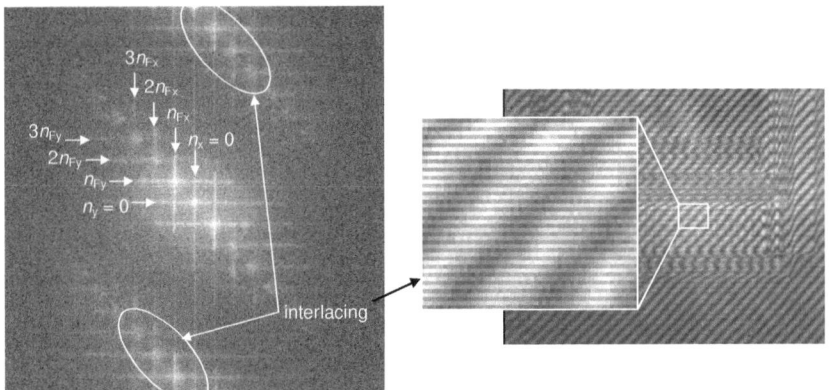

Fig. 2.31. The spectrum of overexposed interferogram with marked higher harmonics. The high frequencies marked by ellipses are caused by the camera interlacing. The interlacing is shown on the right side.

2.5.3 Phase unwrapping

The second procedure of the phase extraction sequence, which is not given only by strict mathematical rules but can be optimised using various unwrapping approaches, is the phase unwrapping (see Fig. 2.18). The unwrapping means removing of the 2π phase steps, which appear after the phase calculation. The unwrapping is very sensitive to the phase steps and noise. If a simple unwrapping method is used, the artifacts introduce an unwrapping error and this can spread into other regions of the phase. Therefore some unwrapping algorithms were investigated.

The following phase extraction algorithms were investigated (see Chapter 1.5): the *straightforward* algorithm, *spiral scanning* algorithm, *pixel queue* algorithm and *minimum spanning tree* algorithm. In the case of the *spiral* and *pixel queue* algorithms the following options were implemented: (i) the number of the neighbouring points, which are taken into 2π-step calculation, was selectable and (ii) the pixels with arcs (see Chapter 1.5.4 for definition) higher than $\pi/2$ were masked and not processed. As a tolerance factor $\beta = 1$ was considered in Eq. 1.22. That means, if the difference between the two neighbouring pixels is $> \pi$ (or $< -\pi$), the phase is corrected by adding (subtracting) 2π.

The invalid unwrapping occurs (i) if a phase step is not abrupt due to the defects (see Fig. 2.32a) or (ii) if the phase difference between two neighbouring pixels is higher than π due to the defects or noise (see Fig. 2.32b). In the first case, the 2π step is not recognised and phase on the right side from the step is not corrected (by subtracting 2π in this particular case). In the second case, an invalid phase step is identified and the phase on the right from the error is shifted by 2π up or down (down in this particular case). The first case is presented on the 2D data example in Fig. 2.32c. An ellipse highlights a non-abrupt phase step. Fig. 2.32d shows the phase after application of a *straightforward* unwrapping algorithm. A wrong step function is assigned to the phase on the right from the highlighted place (dark stripes). More sophisticated unwrapping algorithm has to be therefore used for unwrapping of this region. The phase after the application of the *minimum spanning tree* is shown in Fig. 2.32e. This algorithm detects the points with high error possibility and extracts them at the end. Similar result is obtained using the *pixel queue* algorithm, which overflows the defects and ignores them.

An experimentally measured phase contains many defects and noises like that shown in Fig. 2.32. Invalid unwrapping caused by a single localised phase defect can spread to the whole

2 2D setup

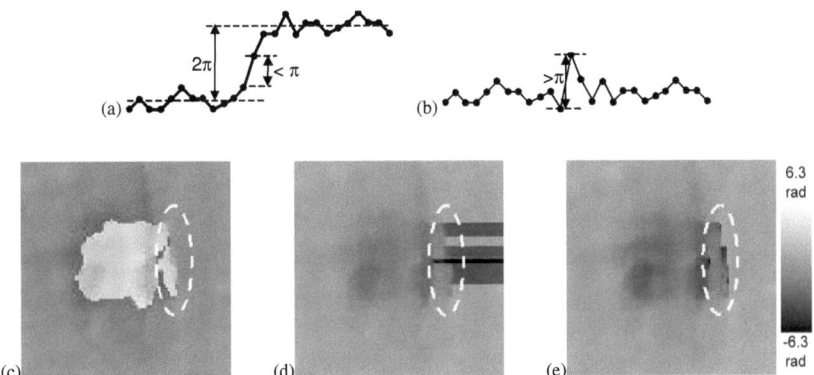

Fig. 2.32. (a, b) Examples of phase defects. (c) 2D phase before unwrapping, (d) phase in (c) after application of a *straightforward* unwrapping algorithm and (e) phase in (c) after application of the *minimum spanning tree* algorithm.

image. Such example is shown in Fig. 2.33. A wrapped phase image after a reference image subtraction is shown in Fig. 2.33a and its cross section along the marked line is in Fig. 2.33e. The 2π steps are positive and negative. Fig. 2.33b shows the phase after application of the *straightforward* linear scan unwrapping algorithm. The noise and defects are interpreted like phase steps. The error of the unwrapping propagates to the right side of the phase image. The phase after application of *minimum spanning tree* algorithm is shown in Fig. 2.33d, the corresponding cross section is in Fig. 2.33e. The phase is unwrapped correctly.

From the example in Fig. 2.33 and from investigation of other phase images, we conclude, that the *minimum spanning tree* and the *pixel queue* algorithms are most suitable for unwrapping of the phases calculated from the interferograms recorded in our 2D TIM setup. However, this high precision is paid by the slow speed. For example, unwrapping of the whole phase image of size around 770x570 pixels using the *minimum spanning tree* algorithm takes seven and half hours in Pentium IV 1.8 GHz 512 MB RAM computer (in *Matlab* environment). The slow speed of these algorithms arises from the complexity of the algorithms. An example of the progressing of the *minimum spanning tree* algorithm is demonstrated in Fig. 2.34. From these reason (slow speed) these two algorithms are only good for unwrapping of local areas and not for unwrapping of the whole phase 770x570 pixels. To avoid the necessity of unwrapping of the whole phase, we developed a phase pre-processing procedure, which is described in following chapter.

2 2D setup

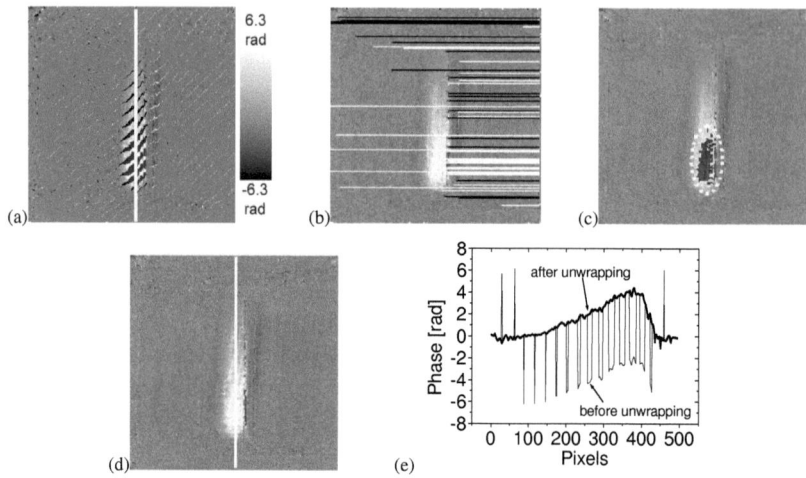

Fig. 2.33. (a) Phase before unwrapping, (b) phase after application of a *straightforward* unwrapping algorithm, (c) phase after pre-processing, (d) phase after application of the *minimum spanning tree* algorithm, (e) cross section along the lines marked in (a) and (d).

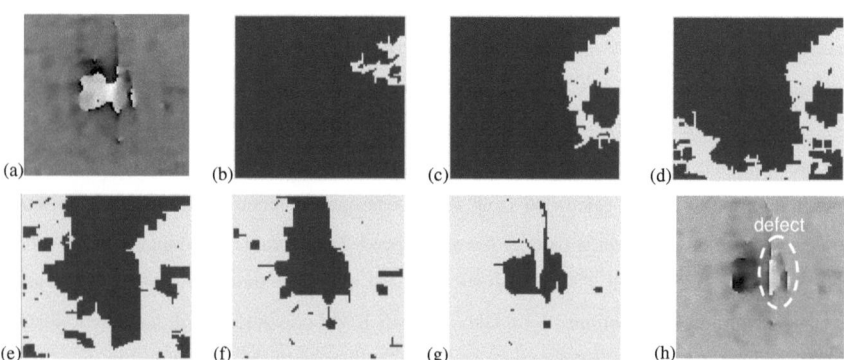

Fig. 2.34. The demonstration of the *minimum spanning tree* algorithm. (a) Wrapped phase, (h) phase after unwrapping. Figures (b-g) demonstrate the progressing after (b) 300, (c) 1000, (d) 2000, (e) 4000, (f) 6000 and (g) 7000 pixels have been unwrapped. They are shown in two colours, where the dark region corresponds to the wrapped phase and light region to the not yet unwrapped phase. In this case the algorithm processes first the points near the edges and only then it proceeds to the middle of the phase, where most of the defects occur.

2 2D setup

2.5.3.1 Phase pre-processing

Before the pre-processing is done, the reference and the stressed wrapped phase have to be subtracted (Step 6 in Fig. 2.18), like it was done e.g. in Fig. 2.33a. This removes the tilt of the phase introduced by the fringe carrier frequency, but the 2π phase steps remain, see Fig. 2.35. The wrapped phase has not a "saw-tooth" profile any more, but a profile of rectangular steps. This is a necessary condition for the pre-processing.

The principle of the phase pre-processing is shown in Fig. 2.36. The whole data set is scanned line-by-line. The intervals with phase value $|\varphi| > \pi$ are detected. The value π has been chosen because it is in the middle between 0 and 2π. We propose, that if the interval begins or ends with the phase step $|\delta\varphi(x_1, x_2)| > \pi$, where x_1 and x_2 are two neighbouring points, all the points in the interval are corrected by subtraction (addition) of 2π. This is the case of the regular phase steps and of the artifacts, see Figs. 2.36a, b. Otherwise it is considered to be a temperature induced phase shift and the interval is not corrected, see Fig. 2.36c. This method is fast (~ 3 s for phase 770x570 pixels) and removes almost all 2π-steps except a part of the heated area, see ellipse in Fig. 2.33c. This small area has to be locally unwrapped with one of the slow unwrapping algorithms. Sometimes all 2π-steps are fully removed and no further unwrapping is necessary.

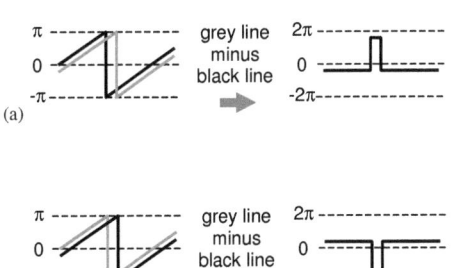

Fig. 2.35. Subtraction of the reference (black line) and stressed (grey line) wrapped phase.

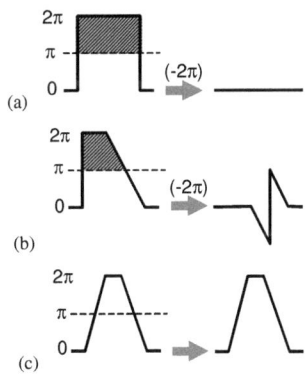

Fig. 2.36. Principle of the phase pre-processing. (a) regular phase step, (b) artifact, (c) temperature-induced phase shift.

2.5.4 Error sources of the phase measurement

In the following part the effect of the laser pulse duration, multiple interference in the DUT, inhomogeneous intensity distribution in the interferogram and instability of the laser pulses on the measurement precision is analysed.

2.5.4.1 Effect of finite pulse duration

The duration of the laser pulse was found to be $2\tau = 5.3$ ns, see Fig. 2.9. During this time the device is illuminated and the interferogram is created. However, the phase stays not constant during this time, see Fig. 2.37b. The interferogram $I(x, y, t)$ is an integration of the interference signal $I'(x, y, t)$ during the laser pulse:

$$I(x,y,t) = \frac{1}{2\tau} \int_{t-\tau}^{t+\tau} I'(x,y,t')W(t')dt' \qquad (2.6)$$

where $W(t)$ is a window function characterising the laser pulse shape in the time scale. In first approximation it is a Gauss function:

$$W(t) = W_0 \exp(-2t^2/\tau^2). \qquad (2.7)$$

The phase can be considered to be linearly increasing in an interval of few nanoseconds, $\varphi(x, y, t) = \phi_0(x, y) + \varphi(x, y, 0) + \partial\varphi(x, y)/\partial t * t$. Here the term $\phi_0(x, y)$ represents modulation of the phase by DUT and the inclination of the reference mirror. The Eq. 1.11 can be rewritten as (see Fig. 2.37b):

$$I(x,y,t) = A(x,y) + B(x,y)\sin\left[\phi_0(x,y) + \varphi(x,y,0) + \frac{\partial\varphi(x,y)}{\partial t}t\right], \quad t \in (-\tau, \tau) \qquad (2.8)$$

Here $t = 0$ corresponds to the middle of the laser pulse, see Fig. 2.37b. Inserting Eqs. 2.7 and 2.8 into Eq. 2.6 and performing the integration lead to:

2 2D setup

$$I(x,y,t=0) = \int_{-\infty}^{\infty} \left\langle A(x,y) + B(x,y)\sin\left[\phi_0(x,y) + \varphi(x,y,t=0) + \frac{\partial\varphi(x,y)}{\partial t}t'\right]\right\rangle * W_0 \exp(-2t'^2/\tau^2)dt' =$$

$$= A(x,y)W_0\tau\sqrt{\pi/2} + B(x,y)W_0\tau\sqrt{\pi/2}\exp\left[-(\frac{\partial\varphi(x,y)}{\partial t}\tau)^2/8\right]\sin[\phi_0(x,y) + \varphi(x,y,t=0)]$$

(2.9)

The amplitude of the fringes in the interferogram is modulated by coefficient $W_0\tau\sqrt{\pi/2}\exp\left[-(\frac{\partial\varphi(x,y)}{\partial t}\tau)^2/8\right]$. If $\partial\varphi/\partial t \neq 0$, the fringe amplitude decreases as a function of $\partial\varphi/\partial t$. Fringe amplitude decreases to $1/e^2$ for phase gradient $\partial\varphi/\partial t = 4/\tau = 4/3.15 = 1.3$ rad/ns. In typical experiments the phase increases with speed approx. $\partial\varphi/\partial t = 0.01$-$0.1$ rad/ns (at the end of 100 ns stress pulse the phase shift is 1-10 rad). This is equivalent to the decrease of the fringe amplitude to $W_0\tau\sqrt{\pi/2}\exp[-(0.1*\tau)^2/8] = W_0\tau\sqrt{\pi/2} * 0.988$ and less. Decrease of the fringe amplitude 0.988-times (i.e. by 1.2 %) can not be recognised in the interferogram. From this we conclude, that the effect of the laser pulse duration on the phase measurement is negligible in our case.

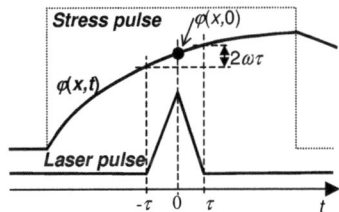

Fig. 2.37. Schematic picture of the laser pulse, stress pulse and phase shift.

2.5.4.2 Multiple reflections in the DUT

The DUT is a structure formed from layers with different refractive index and thickness (see e.g. Fig. 2.14). The incident beam is reflected at every interface between two layers with different refractive index. Depending on the relation between the coherence length L_{coh} of the beam and the distance between the interfaces L, from which the beam is reflected, three basic

2 2D setup

cases are distinguished: $2nL \gg L_{coh}$, $2nL \approx L_{coh}$ and $2nL \ll L_{coh}$, where n is the refractive index of the layer between these two interfaces ($n_{Si} = 3.42$, $n_{SiO2} = 1.46$).

The DUT basically consists of (i) Si substrate, which contains regions with different doping (here the heat source is located) and (ii) SiO_2 layer, which contains the metal layers, see Fig. 2.14. Thickness of the polished Si substrate is usually $L_{Si} = 0.2$-0.4 mm. The beams reflected from the Air-Si interface and Si-SiO_2 interface correspond to the case $2nL \approx L_{coh}$ ($2n_{Si}L_{Si} = 1.4$-2.7 mm). The distances between the Si-SiO_2 interface and the metal layers vary in range of few micrometers. Therefore the beams reflected within the SiO_2 layer correspond to the case $2nL \ll L_{coh}$. For the analysis of the influence of the multiple reflections on the phase shift a fringe simulation software has been developed.

$\underline{2nL \approx L_{coh}}$

For investigation of the effect of the reflections within the Si substrate the structure of Si-layer and metal layer was simulated, see Fig. 2.38a. The thickness of the simulated Si-substrate was 300 μm. The incident beam is E_0, the reflected beams are E_1 (30%[3] energy of incident beam), E_2 (49% energy of incident beam), E_3 (15% energy of incident beam) and E_4 (4% energy of incident beam). Other reflected beams can be neglected in the first approximation due to their low intensity. Beams E_1, E_2, E_3 and E_4 interfere with the reference beam E_R and the interferogram is given by equation:

$$I = |E_1 + E_2 + E_3 + E_4 + E_R|^2 = \sum_{i,j} E_i E_j^* \gamma_{ij} \quad \text{where } i, j \in \{1, 2, 3, 4, R\} \text{ and } E_i = |E_i| e^{i\phi_i} \quad (2.10)$$

where γ_{ij} is the degree of coherence between beams E_i and E_j. The positive phase shift within the Si with a Gauss distribution was simulated, see Fig. 2.38b. The phase shift maximum is 4π. The interferogram was simulated for the coherence lengths 0 μm (see Fig. 2.38c, only beam E_2 interferes with the reference beam), 2000 μm (see Fig. 2.38e, this corresponds to our laser) and 4000 μm (double of our laser). In the simulation we considered that the optical path of the beam E_2 and the reference beam E_R is the same. In Fig. 2.38e the fringe amplitude is modulated due to the multiple reflections. From these interferograms the phase shift is calculated. To see the

---
[3] Reflectivity of interface between layer 1 and layer 2 is calculated by equation $R_{12} = r_{12}^2 = (n_1 - n_2)^2/(n_1 + n_2)^2$.

2 2D setup

difference between the extracted phase shift and the original phase shift, the original and the extracted phase shifts are subtracted from each other. We will call the difference between the extracted phase shift (corresponding to some L_{coh}) and the original phase shift like phase deviation. The phase deviation corresponding to the L_{coh} = 2000 µm is shown in Fig. 2.38d, the phase deviation corresponding to the L_{coh} = 4000 µm is shown in Fig. 2.38f. In the first case (L_{coh} = 2000 µm) the phase deviation varies in range (-0.15, 0.2) rad, in the second case (L_{coh} = 4000 µm) the phase deviation varies in range (-0.5, 0.9) rad. If the Si-thickness is changed within the range of 2 µm, these ranges are changed, but the difference between the maximum and minimum of the range is preserved. From this we conclude, that the maximal possible phase deviation for the structure in Fig. 2.38a can be 0.2 - (-0.15) = 0.35 rad for L_{coh} = 2000 µm and 0.9 - (-0.5) = 1.4 rad for L_{coh} = 4000 µm.

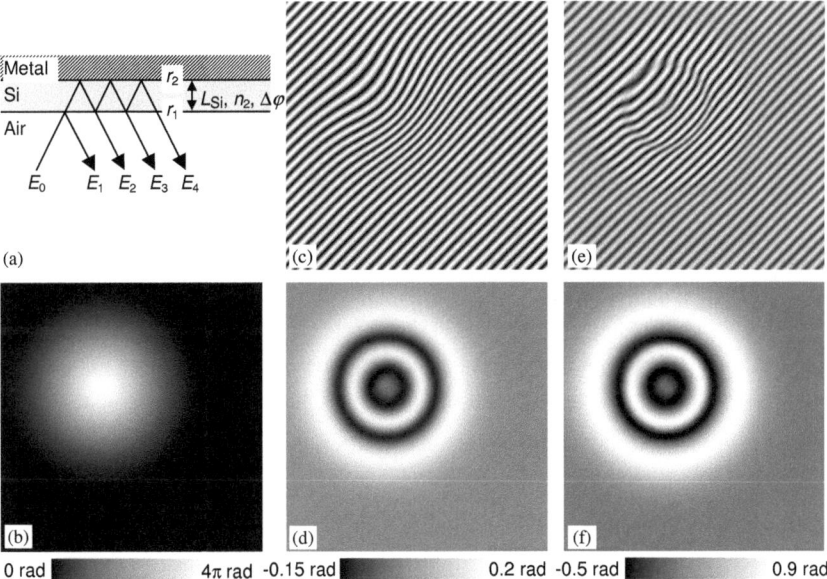

Fig. 2.38. Multiple reflection simulation ($2nL \approx L_{coh}$). (a) The simulated structure Air-Si-Metal and reflected laser beams E_1, E_2, E_3 and E_4. Thickness of Si layer L_{Si} = 300 µm, refractive index n_2 = 3.42. Reflectivity of the metal layer is r_2 = 1. (b) The simulated Gauss phase shift. (c) Simulated interferogram for the coherence length 2000 µm. (d) Difference of the simulated phase shift in (b) and the phase shift extracted from (c). (e) Simulated interferogram for the coherence length 4000 µm. (f) Difference of the simulated phase shift in (b) and the phase shift extracted from (e).

2 2D setup

If instead of the metal layer the SiO$_2$ layer is simulated ($r_2 = 1$ in Fig. 2.38a changes to $r_2 = 0.4016$), the phase deviation is higher. For the $L_{coh} = 2000$ μm the maximal phase deviation can be 0.53 rad and for the $L_{coh} = 4000$ μm the maximal phase deviation can be 3.1 rad.

With thinning of the Si substrate the effect of the multiple reflections increases. For example, if the Si thickness is 200 μm, the phase deviation for $L_{coh} = 2000$ μm increases to 1 rad (Air-Si-Metal) or 1.7 rad (Air-Si-SiO$_2$) and the phase deviation for $L_{coh} = 4000$ μm increases to 1.8 rad (Air-Si-Metal) or even unlimited (Air-Si-SiO$_2$).

The results from above are summarised in Table 2.5, where also data for the Si substrate thickness 400 μm are shown.

Table 2.5. Phase deviation induced by the multiple reflections.

		$L_{coh} = 2000$ μm	$L_{coh} = 4000$ μm
$L_{Si} = 400$ μm	Air-Si-Metal	0.07 rad	0.95 rad
	Air-Si-SiO$_2$	0.11 rad	1.9 rad
$L_{Si} = 300$ μm	Air-Si-Metal	0.35 rad	1.4 rad
	Air-Si-SiO$_2$	0.53 rad	3.1 rad
$L_{Si} = 200$ μm	Air-Si-Metal	1 rad	1.8 rad
	Air-Si-SiO$_2$	1.7 rad	unlimited

In our setup we use the lasers with coherence length $L_{coh} \cong 2000$ μm (see Table 2.2) and the most of the samples have the substrate thickness $L_{Si} = 300\text{-}400$ μm. From this we conclude according to the Table 2.5, that the deviation of the measured phase shift from the real phase shift is less than 0.53 rad. Since deviation by 0.53 rad is the limited case, in most cases the deviation is smaller. The thicker the substrate, the smaller is the deviation.

$\underline{2nL \ll L_{coh}}$

In the case of reflections within the SiO$_2$ layer, the analytical solution for the interference of the multiply reflected beams with the reference beam E_R can be calculated. Let consider a layer structure shown in Fig. 2.39a and a heat $\Delta\varphi$ located inside the Si layer only (this is the most frequent case). The reflections within the Si layer are not considered (they were discussed in the previous part). Because of condition $2nL \ll L_{coh}$, the degree of coherence is equal to one for all beams. The interferogram is given by the following equation:

2 2D setup

$$I = |E_R + E_1 + E_2 + E_3 + ...|^2 = \left| E_R + E_0(1-r_1^2)e^{i(2kL_1n_1+2\Delta\varphi)}\frac{r_2 + r_3 e^{i2kL_2n_2}}{1 + r_2r_3 e^{i2kL_2n_2}} \right|^2 = \quad (2.11)$$

$$= \left| E_R + E'e^{i(2\Delta\varphi+\varphi')} \right|^2 = |E_R|^2 + |E'|^2 + 2|E_R||E'|\cos(2\Delta\varphi+\varphi'-\varphi_R)$$

where:

$$E' = E_0(1-r_1^2)\sqrt{\frac{r_2^2 + r_3^2 + 2r_2r_3\cos(2kL_2n_2)}{1 + r_2^2r_3^2 + 2r_2r_3\cos(2kL_2n_2)}}$$

$$\varphi' = 2kL_1n_1 + \arctan\frac{r_3(1-r_2^2)\sin(2kL_2n_2)}{r_2(1+r_3^2) + r_3(1+r_2^2)\cos(2kL_2n_2)}$$

Here φ_R is the phase of the reference beam. This equation shows that due to the multiple reflections within the SiO$_2$ layer the fringe amplitude and fringe phase is modulated. However the modulation is identical in the reference and in the stressed interferogram and therefore the measurement of the phase shift $2\Delta\varphi$ is not influenced by this effect.

Examples of the interlayer reflections are shown in Figs. 2.39b-e. Fig. 2.39b shows a parasitic fringe pattern highlighted by the oval around the bonding pads. It is caused by the interference of the beams reflected within the passivation layer with varying thickness, see Fig. 2.39c. Figs. 2.39d, e illustrate another case of the interlayer interference. Here the DUT is illuminated by a convergent beam. The interference between the laser beam reflected from the metal and Si-SiO$_2$ interface causes the parasitic fringe pattern highlighted by the oval in Fig. 2.39d.

Antireflection coating

The antireflection coating of the Si backside prevents multiple reflections of the beam within the Si substrate (see Fig. 2.38). The coating decreases the intensities of the multiply reflected beams in Si. The less optical power is distributed into the undesirable beams E_1, E_3, E_4 etc., the more power has the probe beam E_2 (see Fig. 2.38a) and the higher is the accuracy of the phase measurement. Fig. 2.40 compares the interferograms with and without antireflection coating, showing that in the case of sample with the antireflection coating the S/N ratio increases

2 2D setup

from 8 to 13. Correspondingly the standard deviation of the noise in the extracted phase decreases from 0.13 rad to 0.07 rad on the metal.

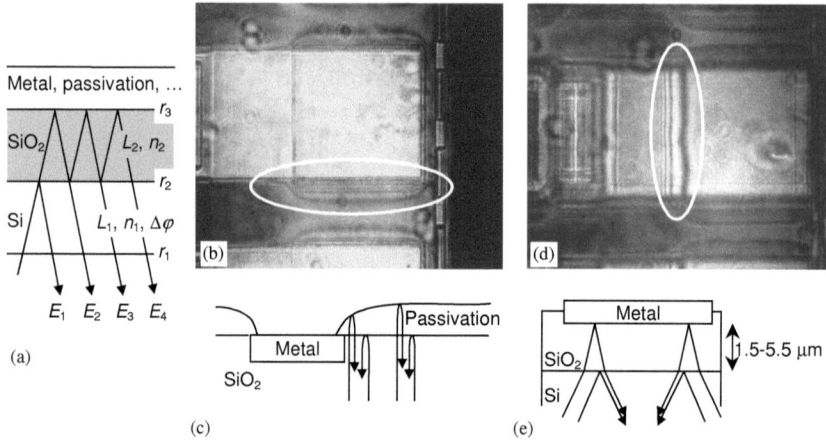

Fig. 2.39. Multiple reflection ($2nL \ll L_{coh}$). (a) Multiple reflections within the SiO_2 layer. (b, c) Metal pad viewed by a parallel beam and the corresponding schematic drawing. (d, e) Metal pad viewed by a convergent beam and the corresponding schematic drawing.

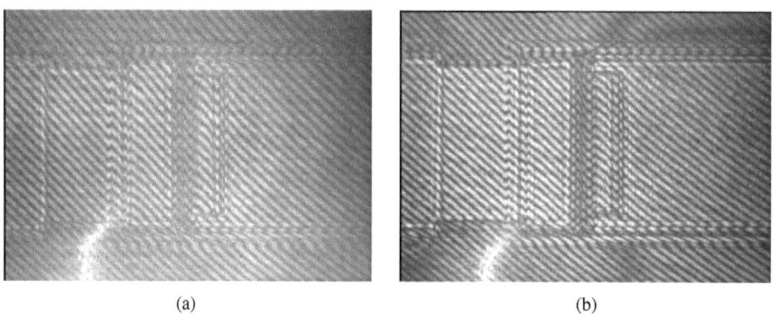

Fig. 2.40. Interferogram of a sample (a) without and (b) with antireflection coating. The S/N on the metal in the (a) case is 8, in the (b) case 13.

2 2D setup

Conclusion

From the results shown above we conclude following for the case of the coherence length of the laser used in the 2D TIM setup ($L_{coh} = 2000$ μm):

- in the case of the Si-thickness $L_{Si} \geq 300$ μm, which is the most case of our experiments, the phase deviation of the extracted phase, caused by the multiple reflections within the device, is comparable with the phase sensitivity $\delta\varphi$ (above the metal and above the SiO$_2$, see Table 2.4). The phase deviation can in a limited case reach the maximal value 0.53 rad. The thicker the substrate, the smaller is this deviation.
- in a case when the Si-thickness is $L_{Si} \sim 200$ μm, the measured phase can differ from the real phase by up to 1.7 rad. This is a value comparable with the usual amplitude of the phase shift and therefore such thin substrates should be avoided.
- the multiple interference within the SiO$_2$ layers does not create any phase deviation.
- to decrease the phase deviation caused by the multiple interference within the Si substrate, a light source with lower coherence length has to be used, or the substrate thickness should be higher, or an antireflection coating has to be applied on the Si backside.

2.5.4.3 *Effect of inhomogeneous laser intensity distribution*

The laser used in the 2D TIM setup is a multimode laser with non-uniform intensity distribution across the beam. This causes variation of the fringe amplitude over the interferogram. The typical experimental interferogram and its cross section are shown in Fig. 2.41. The fringe pattern amplitude ($B(x, y)$ in Eq. 1.11) and offset intensity ($A(x, y)$ in Eq. 1.11) varies along the device cross section. The maximum amplitude is at the centre (S/N = 5) and the minimal is at the corners (S/N = 1).

The maximum of the offset intensity variation is located near the centre of the interferogram spectrum. This part of the spectrum is filtered out during the phase extraction (see Chapter 2.5.2.3) and therefore it has no influence on the phase extraction. The variation of the fringe amplitude is equivalent to the variation caused by the varying reflectivity of the DUT. This effect has influence on the phase and it was discussed in Chapter 2.5.2.2.

2 2D setup

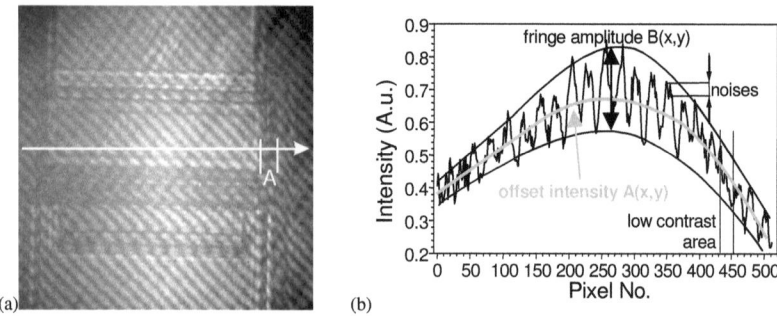

Fig. 2.41. (a) Interferogram and (b) its cross section along the white arrow. The offset intensity variation and fringe amplitude variation due to the inhomogeneous laser beam profile are shown.

2.5.4.4 Pulse to pulse instabilities of the laser beam

The lasers used in the 2D TIM setup exhibit pulse to pulse instabilities of the beam intensity and spatial distribution of the phase front. Due to this the reference and the stressed interferogram may differ in this distribution. This effect can limit the precision of the phase measurement. To demonstrate this effect, the following examples were recorded with the Opolette laser, which exhibits much higher pulse-to-pulse instabilities than Vibrant laser. The reason is that the Opolette laser operates close to the threshold of the light generation, where the generation is less stabile.

An extreme case of the pulse to pulse instability of the beam intensity is shown in Fig. 2.42. The laser light distribution over the field of view changes from pulse to pulse. The simulations of this effect show that the extracted phase shift (obtained by the FFT – based method) is relatively unaffected by such kind of instability.

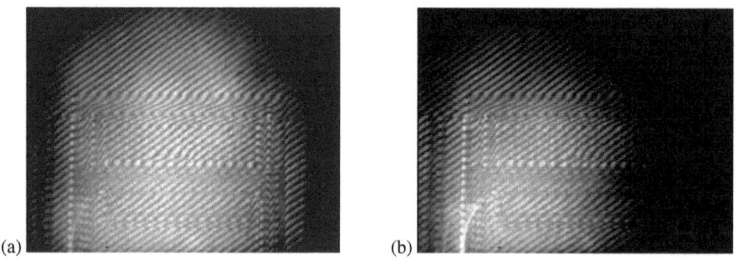

Fig. 2.42. Modulation of the sample interferogram at two pulses due pulse-to-pulse instability of the beam intensity (recorded with the Opolette laser).

2 2D setup

An extreme case of the pulse to pulse instability of the beam phase front is shown in Fig. 2.43. Fig. 2.43a shows the interferogram of a metal pad, Figs. 2.43b-d show the phase difference between two reference interferograms. Since there is no stress-induced phase shift, the source of the phase fluctuations in Fig. 2.43b-d is the instability of the beam wavefront. The fluctuation amplitude reaches the value 0.5 rad in these extreme examples. For the Vibrant laser the fluctuations are usually below 0.2 rad. The fluctuations are superimposed to the measured phase shift, therefore the fluctuations limit the sensitivity of the phase measurement to 0.2 rad in the case of the Vibrant laser.

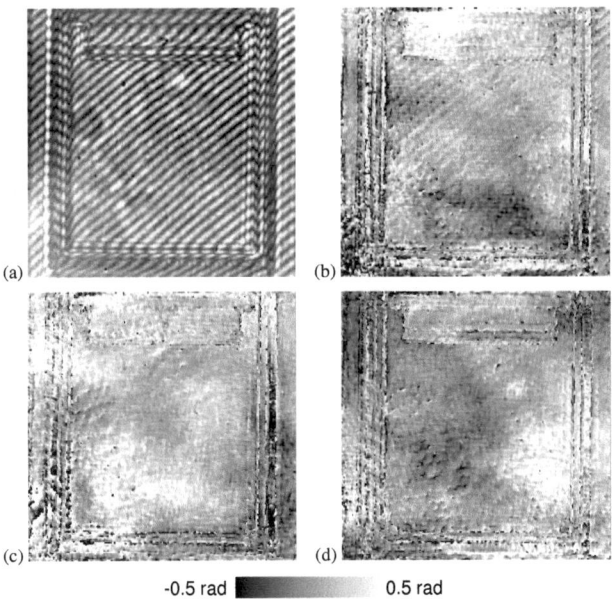

Fig. 2.43. Pulse to pulse instability of the Opolette laser phase front. (a) Interferogram of a metal pad. (b)-(d) Three extreme examples of the phase calculated from two reference interferograms. The phase fluctuates in range (-0.5, 0.5) rad.

2.6 Conclusion – optimal measurement parameters

Based on the previous chapters the following conclusions and optimal parameters for the measurements were found:

1. The field of view has to be chosen in such way that the region of the heat-induced phase shift is approx. 50 % of the interferogram.
2. For rectangular devices, the edges of the investigated device should be parallel to the edges of the interferogram.
3. The optical path between the probe and reference beams should be balanced in a way that the interference image occurs from the device and not from the substrate backside.
4. The reference mirror has to be adjusted in order to choose the optimal fringe thickness (Eq. 2.5). The thinner the fringes, the simpler is the spectrum filtering, but the smaller is the phase sensitivity.
5. The fringe orientation should be approx. 45° with respect to the edges of the interferogram to assure optimal conditions for the spectrum filtering.
6. The constant fringe amplitude over the whole field of view should be ensured for the best result.
7. The optimal settings of the camera sensitivity and laser pulse energy should be engaged to avoid over- and underexposure of the interferogram.
8. Using of the optimal spectrum filter (Fig. 2.30) for the spectrum filtering minimises the undulations and artifact number.
9. The best way to highlight the difference between stressed and reference interferograms is to use the same spectrum filter for their processing and to subtract the corresponding wrapped phases. Next step should be the phase pre-processing (see Chapter 2.5.3.1), which removes almost all 2π phase steps except a part of the heated area.
10. For the local unwrapping the *pixel queue* method or the *minimum spanning tree* method are recommended.
11. Polishing of the sample backside increases the sensitivity.
12. For samples with substrate thickness bellow 300 μm the antireflection coating of the device backside is necessary, otherwise the phase shift will be strongly influenced by the internal interference.

2.7 2D power dissipation density

As was already written in Chapter 1.6 the thermally induced phase shift does not reflect the instantaneous state of the device. To see where the current flows at certain time instant, the 2D power dissipation density P_{2D} has to be calculated by Eq. 1.24. In this chapter we show that in the 2D TIM setup the P_{2D} can be extracted very time-efficiently in comparison with e.g. heterodyne setup.

The space derivatives Ψ of the phase shift φ in Eq. 1.24 can be calculated e.g. using Savitzky-Golay algorithm [Sav64, Pre92], which is applied to the phase recorded in the 2D TIM setup. Savitzky-Golay algorithm is a method for 1D data smoothing and derivative calculation. Distinctive feature of this smoothing is that it can perform satisfactory differentiation of the noisy data.

For the time derivative Σ calculation in Eq. 1.24 a single phase detection is not sufficient. The time derivative term can be approximated using the finite difference between two phases $\varphi(x, y, t)$ and $\varphi(x, y, t-\delta t)$ recorded one after the other during the same stress pulse:

$$\Sigma \approx c_V \frac{\varphi(x,y,t) - \varphi(x,y,t-\delta t)}{\delta t} \qquad (2.12)$$

For this the 2D TIM setup with possibility to record two interferograms at two time instants during a single stress pulse is an optimal tool. The delay δt between the two laser pulses must be as small as possible to keep the correctness of this discretisation. On the other hand, in the case of an experimentally recorded interferogram the noise is present in the phase and small δt leads to a small S/N ratio in P_{2D}, as will be shown later. Therefore an optimal value of δt has to be found.

In the following part we discuss the role of the time (Σ) and space (Ψ) derivative terms in Eq. 1.24 with respect to the thermal diffusion length L_{th} in Si. The thermal diffusion length $L_{th}(t)$ is the distance in which the heat is spread after time t. The thermal diffusion length L_{th} in Si is given by following relation:

$$L_{th}(t) = \sqrt{\frac{t[\text{ns}]}{100}} * 3\mu\text{m} . \qquad (2.13)$$

2 2D setup

We distinguish three basic cases: $2L_{th} < L_{dev}$, $2L_{th} > L_{dev}$ and $2L_{th} \approx L_{dev}$, where L_{dev} is the device size. The coefficient 2 originates from the fact that the heat spreads to the left and right side from the device too.

$2L_{th} \leq L_{dev}$

To estimate the role of the time and space derivative terms if the size of the device is much higher than the thermal diffusion length, a rectangular homogeneous heat source of size 50×100 µm was simulated. For this device the phase shift corresponding to the dissipating power 44 mW/µm² at time 50 ns ($\varphi(t=50ns)$, see Fig. 2.44a) and 80 ns ($\varphi(t=80ns)$) was simulated using the thermal diffusion equation in silicon. The $2L_{th}(80ns) \sim 6$ µm therefore the condition $2L_{th} \ll L_{dev}$ is well fulfilled. The P_{2D} extracted by Eq. 1.24 is shown in Fig. 2.44b. Fig. 2.44c shows the cross section of the phase and of P_{2D}. P_{2D} copies exactly the shape of the simulated heat source.

In Eq. 1.24 the space derivatives can be calculated either from $\varphi(t=50ns)$ or from $\varphi(t=80ns)$. Fig. 2.44c shows the cross section of the time derivative (curve A), space derivative calculated from $\varphi(t=50ns)$ (line B) and space derivative calculated from $\varphi(t=80ns)$ (line C). The difference between the two space derivatives is very small. Using of these two space derivatives in Eq. 1.24 leads to the curves D and E. The average of curves D and E is the actual value of P_{2D}.

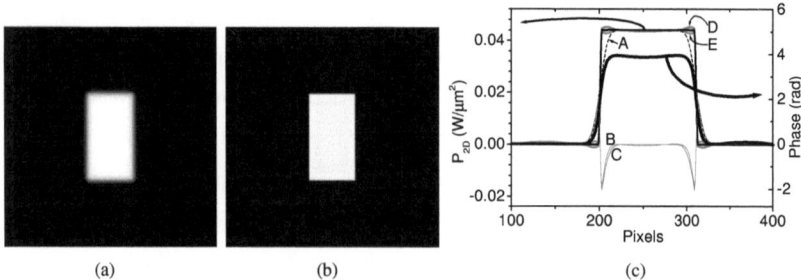

Fig. 2.44. (a) Simulated phase shift corresponding to the heat source of size 50×100 µm and dissipated power 44 mW/µm² at the end of 80 ns stress pulse, (b) extracted P_{2D} and (c) cross section of the phase shift and P_{2D}.

The experimentally measured phase is shown in Fig. 2.45. The space derivative terms (Fig. 2.45d) are much smaller than the time derivative term (Fig. 2.45c) and due to the noise in the phase (Fig. 2.45b) the space derivative terms can be neglected. The P_{2D} (Fig. 2.45e) is thus

2 2D setup

given by the time derivative term. The amplitude of the extracted P_{2D} matches well with the theoretical P_{2D} value of 35 mW/µm², which is calculated from the measured voltage, current and the area of the device. The shape of the extracted P_{2D} fits well to the heat source, except the edges of the heat source, which are relative noisy.

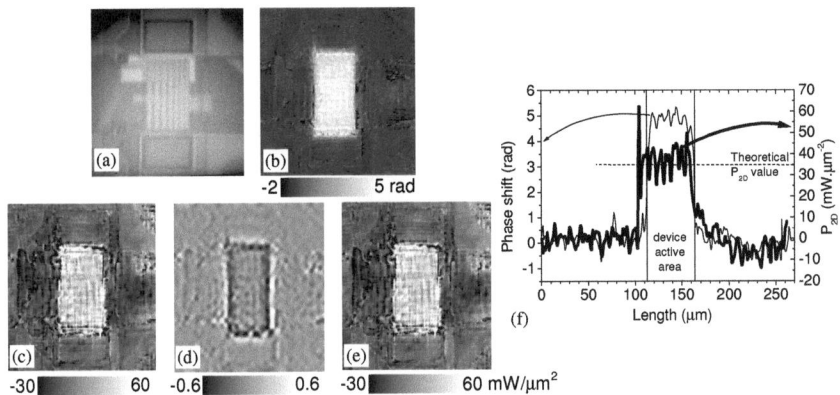

Fig. 2.45. (a) Backside IR image of a resistor of size 50x100 µm. (b) Measured phase shift, (c) time derivative term, (d) space derivative term, (e) P_{2D} and (d) the cross section for the resistor during stress pulse 2.5 A at time 200 ns (dissipated power 35 mW/µm²)

$2L_{th} > L_{dev}$

To estimate the role of the time and space derivative terms if the size of the device is much smaller than the thermal diffusion length, a rectangular homogeneous heat source of size 30x70 µm was simulated. In this case the phase shift corresponding to the dissipating power 2.8 µW/µm² at time 20 µs ($\varphi(t=20\mu s)$, see Fig. 2.46a) and 17 µs ($\varphi(t=17\mu s)$) was simulated. The $2L_{th}(20\mu s) \sim 84$ µm therefore the condition $2L_{th} > L_{dev}$ is again fulfilled, considering $L_{dev} = 30$ µm. The P_{2D} extracted by Eq. 1.24 is shown in Fig. 2.46b. Fig. 2.46c shows the cross section of the phase and the P_{2D}. The shape and size of the heat source can not be recognised from the phase shift, since the heat spreads significantly in such long time scale. However the P_{2D} reproduces nicely the rectangular heat source, see Fig. 2.46b.

Fig. 2.46c shows the cross section of the time derivative (curve A), space derivative calculated from $\varphi(t=20\mu s)$ (line B) and space derivative calculated from $\varphi(t=17\mu s)$ (line C). The difference between the two space derivatives is again very small as well as the difference

2 2D setup

between the corresponding curves D and E, calculated by Eq. 1.24. The average of D and E is the actual value of P_{2D}.

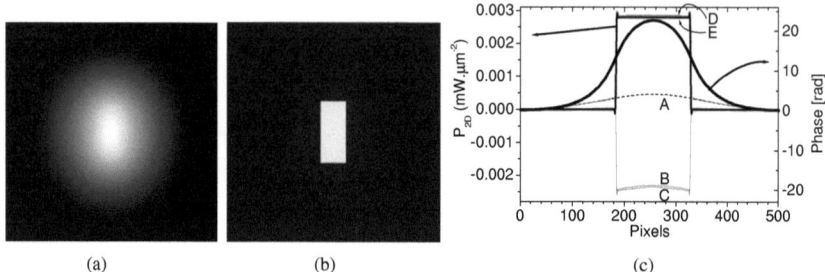

(a)   (b)   (c)

Fig. 2.46. (a) Simulated phase shift corresponding to the heat source of size 30x70 μm and dissipated power 2.8 μW/μm² at the end of 20 μs stress pulse, (b) extracted P_{2D} and (c) cross section of the phase shift and P_{2D}.

The experimentally measured phase is shown in Fig. 2.47. In this case the space derivative terms (Fig. 2.47d) are much higher than the time derivative term (Fig. 2.47c). The time derivative term can be neglected due to the noise in the phase (Fig. 2.47b). The P_{2D} (Fig. 2.47e) is thus given only by the space derivative terms.

The extracted P_{2D} is more than two times higher than the expected value 2.3 mW/μm² (calculated from the measured voltage, current and the area of the device). This is a consequence of the temperature dependence of the thermal conductivity κ in the space derivative terms in Eq. 1.24, which has not been considered (κ= 50-150 WK⁻¹m⁻¹ for temperature range 600-300 K). In the P_{2D} extraction process the value for κ= 150 WK⁻¹m⁻¹ was used and therefore the extracted P_{2D} is higher than the real value.

2 2D setup

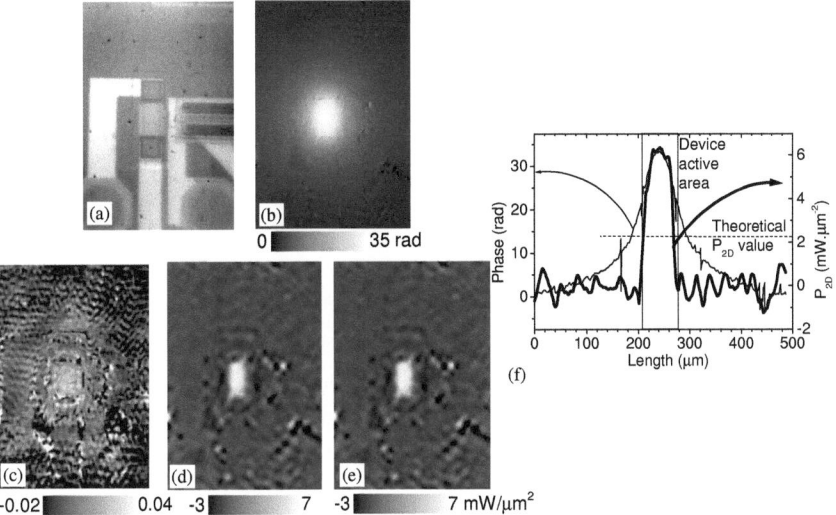

Fig. 2.47. (a) Backside IR image of a resistor of size 30x70 μm. (b) Measured phase shift, (c) time derivative term, (d) space derivative term, (e) P_{2D} and (d) the cross section for the resistor during stress pulse 0.35 A at time 200 μs (dissipated power 2.3 mW/μm²).

$\underline{2L_{th} \approx L_{dev}}$

In previous two parts it was shown, that for $2L_{th} < L_{dev}$ the space derivative terms Ψ in Eq. 1.24 are much smaller than the time derivative term Σ and can be neglected due to the noise. For $2L_{th} > L_{dev}$ it is vice versa and the time derivative term Σ can be neglected. These are two extreme cases. In the case $2L_{th} \approx L_{dev}$ all derivative terms are comparable and to extract the P_{2D} all terms have to be calculated.

The experimentally obtained sensitivity of the P_{2D} measurement is summarised in Table 2.6.

Table 2.6. P_{2D} measurements sensitivity.

Stress pulse length (μs)	P_{2D} sensitivity estimation (mW/μm²)
0.2 - 2	20
2 - 20	4
20 - 200	2

2.7.1 Optimal delay δt

There is no exact criterion for choosing an optimal delay between the two time instants δt in Eq. 2.12. On one hand the delay has to be as short as possible to keep the approximation of Eq. 2.12. On the other hand it cannot be too short since the noise in P_{2D} would be higher than the signal. The optimal value of δt depends on the dissipated power, time scale, noise and speed of current filament.

In Fig. 2.48 the P_{2D} in an ESD protection device at time 200 ns for different delay δt is calculated. For $\delta t = 10$ ns a current filament is recognised but its size and amplitude is not possible to measure due to the noise. The optimal delay has been found to be $\delta t = 30$ ns in this case, since the filament size and amplitude are clearly recognised. If a longer delay was used, the calculated filament size would be larger than the real size.

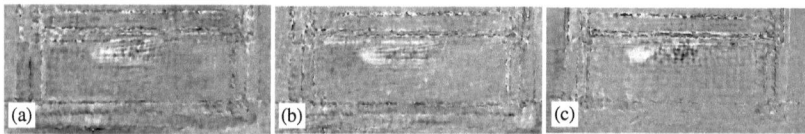

Fig. 2.48. P_{2D} in ESD protection device calculated for different delays δt between the two laser beams: (a) 10 ns delay, (b) 20 ns delay and (c) 30 ns delay. All three images are shown in the same scale (-30, 30) mW/μm².

2.8 Examples

2.8.1 Example 1 – Phase calibration

To compare the phase measurement precision of the 2D TIM setup with the heterodyne scanning setup, a diode structure was investigated both setups under the same conditions, see Fig. 2.49a. The device was stressed with current pulses of amplitude 0.75 A and duration 100 ns. The interferogram of the device, the phase shift at the end of the stress pulse and the comparison of the results from 2D TIM setup and heterodyne setup are in Fig. 2.49. The phase shift from the 2D TIM setup is noisier and with undulations, but quantitatively the results match very well. The undulations were discussed in Chapter 2.5.2.

2 2D setup

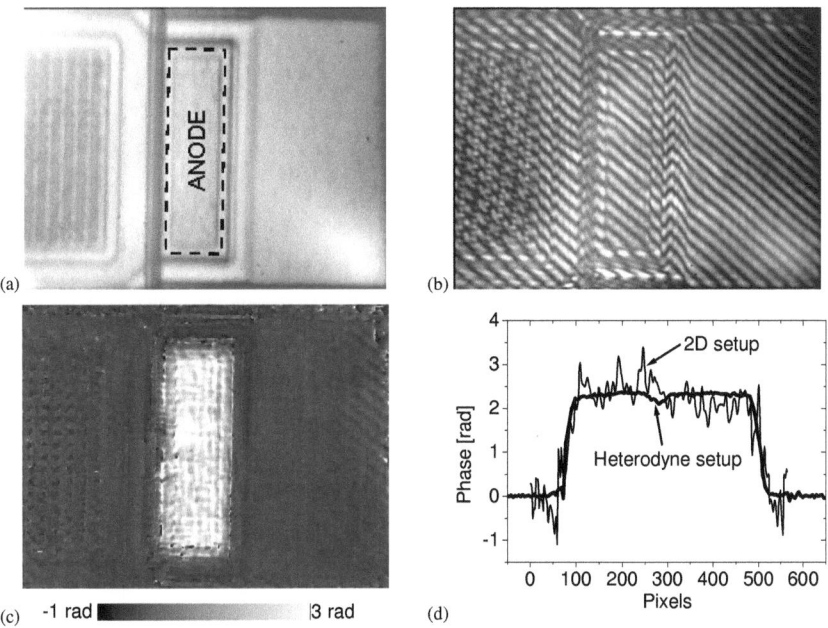

Fig. 2.49. Calibration of the 2D TIM setup. (a) Backside IR image of the diode device, the anode size is 30x100 µm. (b) Interferogram of the device before the stress pulse, (c) extracted temperature induced phase shift, (d) vertical cross section of the phase shift along the device width (thin line) in comparison with the phase shift measured in the heterodyne scanning setup (thick line).

2.8.2 Example 2 – Moving current filament

A coupled npn/pnp ESD protection device was investigated in the 2D TIM setup with the goal to detect the current filament movement [Dub04b]. The device backside IR image is in Fig. 2.50a. It is a combination of vertical npn and lateral pnp transistor, see cross section in Fig. 2.50f. Device deep buried layer serves as a collector of npn transistors. The buried layer is connected by an n-sinker to the silicon surface. In the breakdown a voltage drop in the p-body is created due to the internal p-body resistance and the vertical transistor is triggered into the snap back. The lateral pnp transistor opens at currents higher then used in this experiment.

The measured phase shifts and the corresponding P_{2D} at four time instants are shown in Fig. 2.50. In order to calculate the P_{2D} the second laser beam was placed 30 ns after the first one.

2 2D setup

Due to the negative differential resistance [Esm00] of the device in snap back, a current filament (represented by a spatially localised P_{2D}) is formed in the middle of the device at the pulse beginning, see Fig. 2.50g. Due to the negative temperature dependence of the impact ionisation rates, the current filament starts to move into a cooler region along the device width [Den04, Pog05]. The direction of the movement to the left or to the right is random from pulse to pulse. In case shown in Fig. 2.50 it moves to the left corner. When the filament reaches the corner of the device, it is "reflected" back (Fig. 2.50i) and moves again to the middle and toward the opposite corner of the device (Fig. 2.50j). The current filament amplitude is about 30 mW/µm^2, which is consistent with the results obtained previously using the time-consuming scanning method [Dub04b, Pog05].

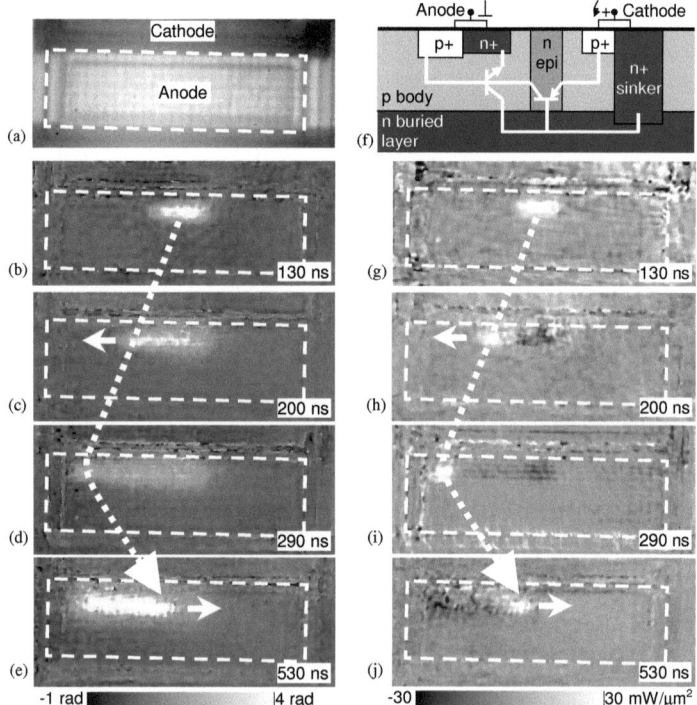

Fig. 2.50. (a) Backside IR image of the ESD PD and (f) device cross section. The anode area is approximately 40x130 µm. (b)-(e) Phase shift distribution $\varphi(x,y,t)$ in the ESD PD at four time instants during current pulses of 0.25 A, 600 ns duration. The heating causes a positive phase shift, which is displayed by light colour. (g)-(j) show the P_{2D}, representing the current filament position. The dashed arrow and the small white arrows denote the movement direction.

2.8.3 Example 3 – Unrepeatable device triggering in power DMOS

The P_{2D} in a self-protecting vertical DMOS transistor with 440 cells was measured in order to understand the current flow distribution in snap-back [Dub04b]. The backside IR image of the DMOS device is shown in Fig. 2.51a and its cross section in Fig. 2.51b. Figs. 2.51c and 2.51d show a sequence of images for two different single pulses at two time instants. At snap back, when the parasitic bipolar transistor of the DMOS turns on, a few cells take over the current and localised current filaments are formed [Den03, Den04]. The position and number of the filaments vary randomly and non-repeatedly from pulse to pulse, as seen by comparing Fig. 2.51c and 2.51d. Shortly after the pulse beginning (see images at $t = 60$ ns), the filaments are formed along the source field termination [Den04]. Five and three filaments are active in the case of Fig. 2.51c and 2.51d, respectively. With time, the filaments penetrate into cooler regions either along the termination or inside the device (see Figs. 2.51c,d at $t = 120$ ns and 200 ns, respectively). In the example Fig. 2.51c, one of the filaments (marked by circle "A") splits into two branches, whereas another filament (marked by circle "B") neither moves with time, nor the temperature in it increases. This means that the current in it is negligible at later times.

In order to extract the P_{2D} in the DMOS device, the two imaging laser beams were placed at time instants 120 ns and 150 ns, see the corresponding phase shifts in Fig. 2.52 on the left side. The P_{2D} calculated corresponding to $t = 150$ ns is shown on the right in Fig. 2.52. At this instant, DMOS cells inside of the source field are active (i.e. carry current). Some particularities can be observed. The filament marked by circle 'A' is probably a single filament, which later splits into two filaments. The filament 'B' as well as the filaments in 'A' are localised in few DMOS cells. In some case we also observe that the filament extends or splits in a relatively large area as for example the filament in circle 'C'.

2　2D setup

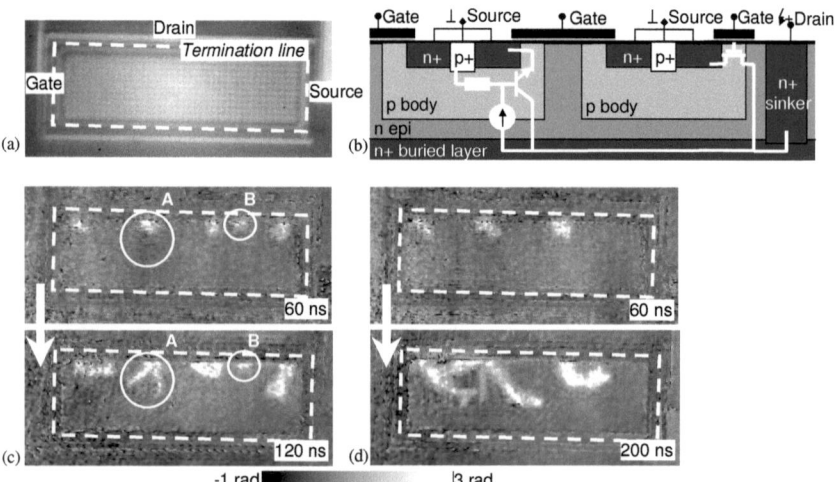

Fig. 2.51. Moving filaments in the vertical DMOS structure during a 3 A, 200 ns stress pulse. (a) Backside IR image of the DMOS, (b) cross section of two DMOS cells with marked DMOS transistor and parasitic vertical bipolar transistor. Two examples of triggering behaviour are shown: (c) five filaments at 60 ns and 120 ns after the pulse beginning, (d) three filaments at 60 ns and 200 ns after the pulse beginning.

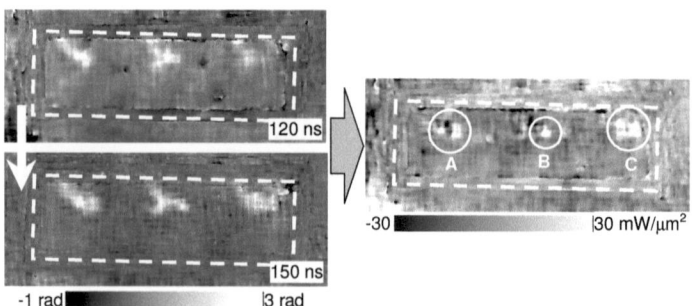

Fig. 2.52. Phase shift distribution in the vertical DMOS at 120 ns and 150 ns during a 3 A, 200 ns stress pulse (left side) and P_{2D} calculated from this two phase shifts (right side). The current filaments are highlighted by circles.

The advantage of the P_{2D} measurement compared to the one-time-instant 2D TIM method or a heterodyne setup is that it allows investigation of unrepeatable effects. Here one clearly finds (i) whether a specific filament stays active or not during the stress, (ii) where the active filament moves and (iii) what is the approximate extension of the current filament.

2.8.4 Example 4 – Destructive phenomena measurement

The main reason of the failure in power and ESD protection devices is the thermal destruction due to the second breakdown [Dwy90, Ame93]. The 2D simulation cannot accurately predict the failure level and pathways due to inaccurate physical models at high temperatures and 3D effects. The current filaments are possible to measure in the 2D TIM setup. Here an example of the destructive process is analysed in the vertical DMOS structure, which was introduced in the previous example.

The DMOS structure was stressed with a series of constant current pulses and after each series the stress current was increased. After applying of six stress pulses to the device with the stress amplitude 3 A the damage occurred. In Fig. 2.53a the voltage and current waveforms of the destructive pulse are shown. The voltage on the device drops at time approx. 160 ns from the pulse beginning (arrow A), when the device goes to second breakdown. First laser pulse illuminated the device before the second breakdown (at 60 ns), second laser beam during the second breakdown (210 ns). Corresponding phase shifts are shown in Fig. 2.53b,c, respectively. In Fig. 2.53c the destructive current filament is marked in a circle. The temperature (phase shift) in this point increases enormously, since all the current flows in a single localised area. The filament formation results in a permanent damage with the same size as the filament. The damage can be seen as a dark spot in the device backside IR image in Fig. 2.53d. Every next stress pulse drives the DMOS into second breakdown represented by a current filament at the same place. Similar effect but in different structure were published in [Pog03b].

2 2D setup

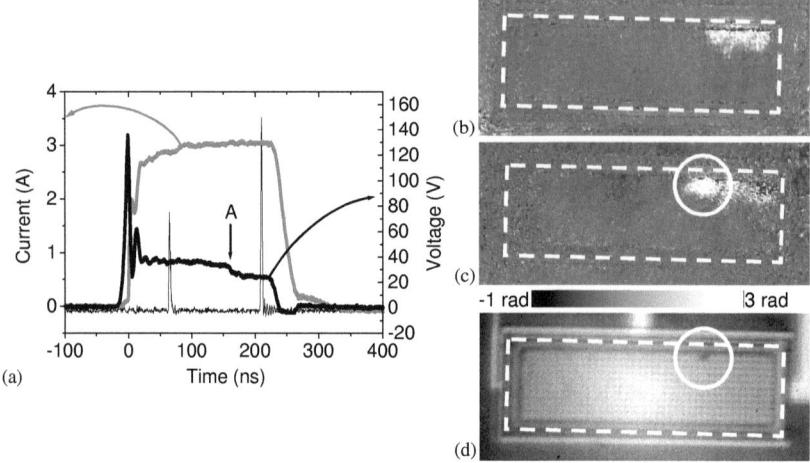

Fig. 2.53. Destructive stress pulse (3 A, 250 ns). (a) Current and voltage waveform of the stress pulse and the two laser-beams positions. (b) Phase shift at time 60 ns and (c) at 210 ns, (d) the backside IR image with marked position of the damage.

2.9 Thermal imaging using absorption measurements

Another option for the nanosecond imaging of temperature and free carrier concentration changes in the DUT under the electrical stress pulse is the measurement of the light absorption in the semiconductor bulk in the 2D TIM setup [Pog03a]. For this the reference branch of the setup (see Fig. 2.1a) is blocked and only the probe beam creates an image on the camera. Thus any interference is avoided.

As was already described in Chapter 1.4.2, the absorption coefficient locally increases with increasing temperature or free carrier concentration in the silicon substrate. The intensity of the reflected light therefore decreases. During the experiment a reference image $I_r(x, y)$ is recorded before the stress pulse is applied to the device and a stressed image $I_s(x, y)$ is recorded during the stress pulse, as well as in the interferometric measurement. Using the Eq. 1.6, the normalised intensity change of the reflected light beam $\Delta I / I_r$ at a lateral position (x, y) can be expressed as follows:

2 2D setup

$$\frac{\Delta I}{I_r}(x,y) = \frac{I_r(x,y) - I_s(x,y)}{I_r(x,y)} = 1 - \chi(x,y)\exp\left[-2\int_0^L \Delta\alpha(x,y,z)dz\right] \quad (2.14)$$

where $\Delta\alpha = \alpha_s - \alpha_r$ is the change in the absorption coefficient introduced by the electrical stress pulse. α_r and α_s are the absorption coefficients before the stress pulse and during the stress pulse at certain time instant, respectively. The integration in Eq. 2.14 is taken along the laser beam path, where L is the substrate thickness. The coefficient χ is equal to 1 for a laser source without pulse-to-pulse instability. For the laser source with pulse instability, like the lasers used in the 2D TIM setup, coefficient χ is not a constant and it is a function of lateral position (x, y).

Since the absorption depends on the light wavelength and the temperature, correct selection of the wavelength is important, see Fig. 1.3b. If the ambient temperature during the experiment is around 300 K, the highest change of the absorption takes place for wavelengths around 1264 nm according to the Fig. 1.3b. Using of a light source with wavelength around 1264 nm therefore leads to the best measurement sensitivity. For higher or smaller wavelengths the change of the absorption with temperature is less relevant (gradient of the curve is smaller) and the measurement sensitivity decreases. Additionally, for smaller wavelength (around 1164 nm) the absorption of light in silicon at ambient temperature is high and the quality of the images therefore decreases resulting into even smaller sensitivity.

From Fig. 1.3b it can be seen that the absorption coefficient does not change linearly with the temperature. If e.g. the wavelength 1264 nm is used, the absorption increases rapidly during first 100 K and then it increases just slowly with increasing temperature. This is different from the interferometric method, which shows a monotonic increase of the measured phase shift with temperature. Consequently the dynamic range of the absorption imaging method has an upper limit, in comparison with the interferometric method.

2.9.1 Example – Spreading current filament

The absorption measurement technique is used here to study the current flow distributions in a npn bipolar transistor with short circuited base/emitter region and a buried collector [Pog02b]. Figs. 2.54a and 2.54b show typical single-shot images of the sample before (I_r) and during (I_s) the stress pulse, respectively. The reflectivity image in Fig. 2.54b was

2   2D setup

recorded at time 50 ns after the pulse beginning. A dark region due to temperature-induced increase of absorption can be observed in half of the device (see the parentheses). This indicates an inhomogeneous current flow in the device shortly after the beginning of the stress pulse.

In order to highlight the temperature-induced absorption, the intensity change $\Delta I = I_r - I_s$ is plotted. Figs. 2.54c and 2.54d show ΔI of the device at time 50 ns and at time 150 ns after the pulse beginning, respectively. Compared to the image at 50ns, at 150ns the absorptive region spreads over the whole device width. This demonstrates that the heated region (current filament) spreads out with time during the stress pulse.

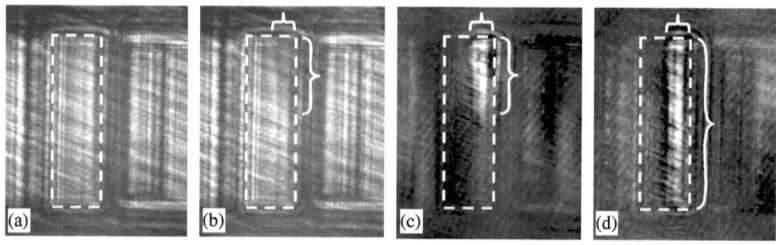

Fig. 2.54. Absorption in ESD protection device. Dashed rectangle indicates the anode area. (a) Reference image of device (before the stress pulse was applied), (b) image of device at time 50 ns after the beginning of 3 A stress pulse. The parentheses indicate the position of the hot region. (c) Subtraction ΔI of images (a) and (b). Light colour indicates region of absorption. (d) Subtraction of reference image (a) and image recorded at time 150 ns.

2.10   Setup further development

There are several further improvements of the 2D TIM setup which are possible in the future, but limited by the state-of the art technology. The most relevant is the replacement of the expensive high power OPO laser by a cheaper and more compact laser source. The best candidates are pulsed semiconductor lasers and fiber lasers. Another improvement is the use of InGaAs IR focal plane array camera with high sensitivity in near IR region.

At the present time two high power lasers are used for creating images. If a much simpler pulsed laser diode could be used for the imaging, the system would be more compact, simplified, the maintenance would be simpler and the pulse to pulse instability would be avoided. There would be no need for the exchange of the water deionising filter and laser system realignment.

2 2D setup

As well the setup cost would be reduced. However, the state-of-the-art power semiconductor pulsed laser diodes exhibit several order of magnitude lower power compared the presently used Q-switch lasers. Therefore a more sensitive camera detector would be needed. Focal plane array (FPA) camera based on InGaAs in combination with a high power pulsed laser diode represents one possible alternative. It was experimentally confirmed that the sensitivity of the FPA camera is 2-3 orders of magnitude higher than that of the presently used vidicon camera. The estimation based on our tests shows that the peak power of a semiconductor laser diode should be at least 20-50 W (100-250 nJ for 5 ns pulse width, see Fig. 2.55) in order to obtain images with a sufficient signal to noise ratio in FPA camera. However, the power of a best available single element pulse laser diode (Perkin&Elmers) is not yet sufficient (e.g. max 4 W) for imaging with 5 nanosecond time resolution. There exist laser diode modules with power as high as several tens of Watts but such systems are composed of several parallel laser diodes. The disadvantage of a parallel diode combination is the lost of the space coherence of the laser source (only the light emitted by a single element is coherent). This is necessary to achieve the interference contrast. It is expected that in few years such a single emitter diode with enough power will be available on the market.

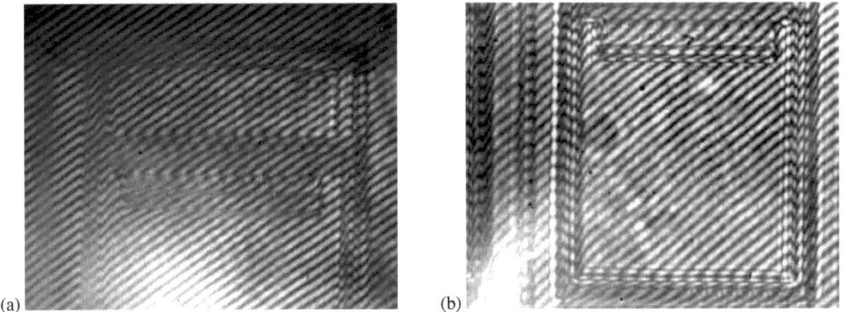

Fig. 2.55. Interferograms recorded in FPA camera using different laser sources: (a) pulsed laser diode (wavelength 1300 nm, peak power on the camera chip 0.22 mW, pulse duration 10 μs), (b) Opolette laser (wavelength 1300 nm, peak power on the camera chip 3.6 W, pulse duration 5 ns). The laser beam is approx. 5 times attenuated by the setup.

The imaging using a presently available semiconductor laser diode and a FPA camera would be possible with time resolution 50-100 ns or longer. Such a resolution is not suitable for the analysis of ESD phenomena occurring in the 100 ns time scale. The time resolution is

2 2D setup

however suitable for imaging of devices subjected to long pulses (µs to ms range), such as power DMOS in short-circuit operation.

Another option is to use a high power pulsed fiber source e.g. produced by OZ Optics. This fiber laser provides a single mode beam with peak power up to 20 kW at wavelength 1550 nm and has a short coherence length 0.6 mm. The only disadvantage of this laser is that it is a high repetition rate (operates in kHz-rates) and an addition shutter would be necessary to install into the setup.

Next suitable laser source is a diode pumped solid state (DPSS) laser, where a laser crystal is pumped by light of an LED array. Such a system is more compact as the used OPO system. At the present, the disadvantage of the available DPSS lasers is either a long laser pulse length or a passive Q-switch.

By selecting of the correct laser source a short coherence length (< 2 mm in the air) and single spatial mode operation should be assured.

3 Dual-beam interferometer

3.1 Introduction

For investigation of semiconductor devices in the CDM time domain a setup with sub-nanosecond time resolution is needed. The time resolution of the 2D TIM setup is limited by duration of the laser pulse to 5 ns (see Chapter 2.3.4). The time resolution of the heterodyne setup is limited by the detector bandwidth to few nanoseconds [Lit03]. This is a trade-off between the bandwidth and S/N ratio. The differential interferometers [Fur99, Dil97, Dil99] provide information only on the difference phase signal, therefore some important device behaviour may be hidden or interpretation of the measurement is not straightforward. Therefore a dual-beam Michelson interferometer (MI) has been developed with the time resolution of 0.4 ns. In this setup the absolute phase shift caused by the refractive index changes at two different positions is simultaneously measured. This allows sub-nanosecond investigation of device behaviour (e.g. trigger instabilities, trigger delays) during a single stress pulse, which is not possible by a single beam setup.

3.2 Setup description

In this chapter the dual-beam interferometer is introduced, the setup parameters are shown and timing of the setup is described. This setup is an extension of single-beam setup also developed within this work.

3.2.1 Optical layout

The dual-beam interferometer consists of two independent MIs combined into one setup. The setup schematic is depicted in Fig. 3.1a, the setup photograph is in Fig. 3.2. Two laser diodes of wavelength 1.3 µm with distributed feedback (DFB) are used here. Due to the DFB the coherence length of the diodes is several hundreds of millimeters. In order to reduce the optical

3 Dual-beam interferometer

losses when combining these two probe beams, the beams have an orthogonal polarisation and are combined using a polarizing cube beam splitter PBS (see Fig. 3.1a). The orthogonal polarisation is achieved by using of a $\lambda/2$ wave plate in combination with cube polarizer (see Attenuator in Fig. 3.1a). The two beams are focused by the microscope objective on the DUT (see also Fig. 3.1b), which is mounted on x-y-z and rotational stage. Each reference branch is piezo-controlled for the adjustment of the working point. One of the beams has a fixed position, while the position of the second beam can be adjusted by mirror 3. The position of the beam spots on the device can be visualised using an infrared vidicon camera and the broadband light source. The interference signals related to two interferometers are detected by two InGaAs detectors (New Focus 1611-AC). The detectors, current and voltage waveforms are sampled with a 4 GHz digital oscilloscope (Tektronix TDS 7404).

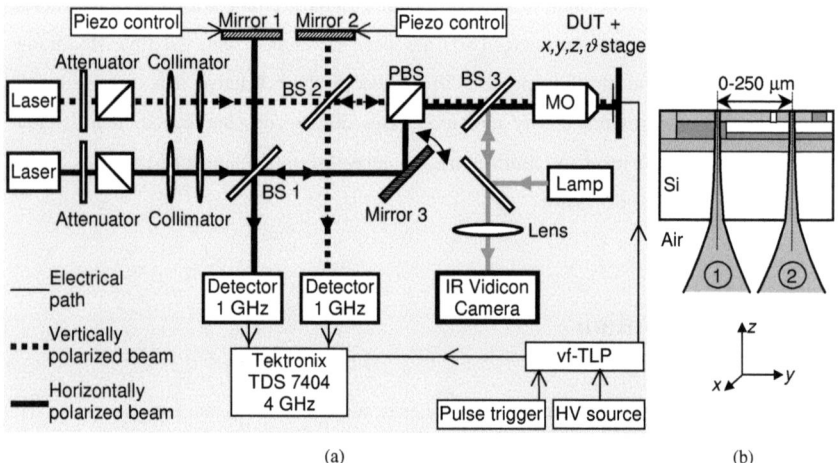

Fig. 3.1. (a) Simplified schematic of the dual-beam Michelson interferometer setup: BS- beamsplitter, PBS- polarisation beamsplitter, MO- microscope objective, DUT- device under test. (b) Focusing of the two laser beams.

3 Dual-beam interferometer

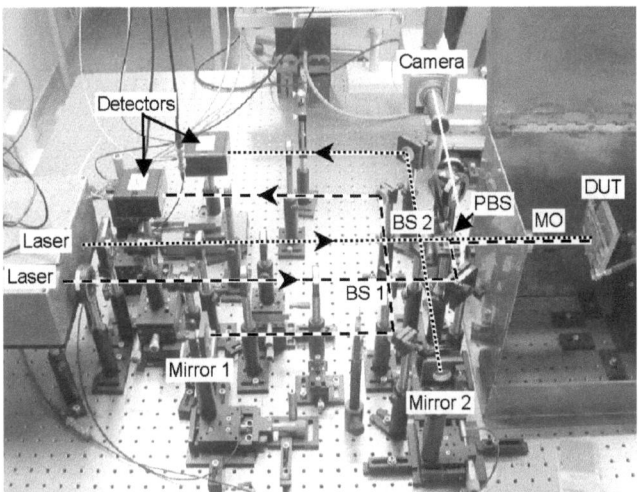

Fig. 3.2. Picture of the dual-beam interferometer.

3.2.2 Setup parameters

The beam spot diameter is determined by the objective magnification, see Table 3.1. For the objective with 50x magnification the beam spot size in the DUT is approx. 2.3 µm. This spot size determines the spatial resolution of the setup. In Table 3.1 also other parameters are shown like the field of view and maximal beam separation, which is described in Fig. 3.1b.

Table 3.1. Dual-beam interferometer parameters.

Objective	Mitutoyo 10x	Mitutoyo 20x	Mitutoyo 50x
Field of view (µm)	1050 x 810	520 x 390	210 x 160
Pixel size (µm)	1.4	0.6806	0.2786
Beam spot diameter FWHM (µm)	14	6.1	2.3
Maximal beam separation* (µm)	230	110	34

3 Dual-beam interferometer

The time resolution of the setup is limited by the rise time of the detectors to 0.4 ns (detectors bandwidth is 1 GHz). The detector with this bandwidth was chosen as a trade-off between the acceptable S/N ratio and the speed.

The phase sensitivity is better than 0.1 rad for the single pulse measurement, depending on the reflectivity of the device (see Table 2.3). For devices without pulse to pulse instability in current flow the signal can be averaged during several pulses to improve the S/N ratio.

3.2.3 Electrical device testing

For characterisation of the device behaviour under the ESD the repetitive pulsing is necessary. The repetitive pulsing is achieved by pulsers based on transmission line pulse (TLP) technique [Mal85, Gier99]. Two TLP generators are used in this work: the TLP system for emulation of the HBM ESD and the very fast TLP (vf-TLP) system for emulation of the CDM ESD.

3.2.3.1 TLP technique

The TLP technique is applied to generate the current pulses of length 100-150 ns for investigation in the HBM time domain. Its scheme is shown in Fig. 3.3. A coaxial cable (transmission line – TL) is charged to a high voltage (HV) V_{charge} and then discharged by closing a reed relay (SW). By this a rectangular voltage pulse is generated, which length $\tau_{TLP} = 2L_{TLP}/v_{coax}$ depends on the length L_{TLP} of the cable (TL) and speed of the electromagnetic wave in the cable v_{coax}. To create a current pulse a 1 kΩ resistor is put in series with the DUT, which impedance is usually much smaller. A 50 Ω resistor placed parallel to this system to avoid the back-reflections. A current probe (CP) and voltage probe (VP) measures the current through the DUT and the voltage on the DUT, respectively. The current I_{DUT} through the device is following:

$$I_{DUT} = \frac{1}{2}\frac{V_{charge}}{1\,k\Omega}. \tag{3.1}$$

3 Dual-beam interferometer

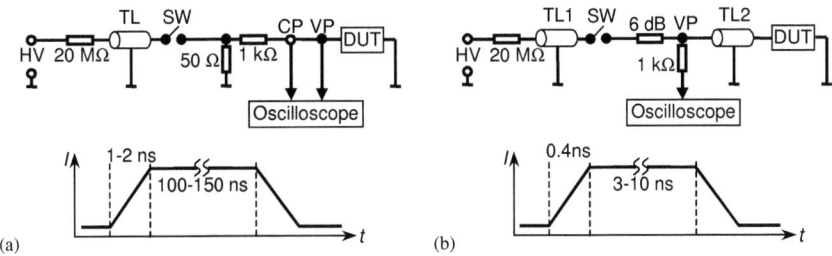

Fig. 3.3. The diagram of the (a) TLP and (b) very fast TLP pulser. Bellow each diagram a typical pulse waveform and time intervals are illustrated.

The presented TLP pulser is not usable in 2D TIM setup, since the reed relay SW has a jitter in the range of 0.1-1 microsecond that disables the synchronisation between the electrical and optical pulse.

3.2.3.2 vf-TLP technique

For investigation in the CDM time domain, the coaxial cable TL1 is discharged through the transmission line TL2 directly to the device, see Fig. 3.3b. This method is based on the time domain reflectometry (TDR) [Gie99]. The voltage probe VP measures the incident voltage waveform (amplitude V_i), coming from the coaxial cable TL1, and the waveform reflected from the DUT (amplitude V_r). The line TL2 has to be long enough to avoid overlapping of these two waveforms in the oscilloscope. The voltage and current waveforms on the DUT can be then calculated [Gie99]:

$$V_{DUT} = V_i + V_r$$
$$I_{DUT} = \frac{V_i - V_r}{50\,\Omega} \tag{3.2}$$

The risetime of the pulses is 400 ps. This method is called very-fast TLP (vf-TLP) [Gie98] to distinguish from the current source TLP.

3.3 Phase measurement and calculation

In this chapter we show how the phase shift is calculated from the detector signal. The dual-beam setup is designed for the measurement of small phase signals bellow π. Therefore also the phase calculation shown bellow is valid only for such small phase shifts. In the case of phase shifts higher than π the one-dimensional unwrapping would be necessary similar to that shown in the Chapter 2.5.3.

The two laser beams are focused in two fixed points (x_1, y_1) and (x_2, y_2) on the DUT. The signal in these points is a function of time t and the Eq. 1.11 can be written in form:

$$I(t) = A + B\cos[\phi(t)] = A + B\sin[\varphi_0 + \Delta\varphi(t)], \qquad (3.3)$$

where φ_0 is called working point. Before the measurement starts, the working point φ_0 is specified by the phase of the reference beam (i.e. position of the reference mirror). For the measurement of small phase shift ($< \pi/2$), the working point should be placed in the region of maximal sensitivity. This is either in the zero-crossing point of the sinusoid (see Fig. 3.4a), or near to the middle (see Fig. 3.4b).

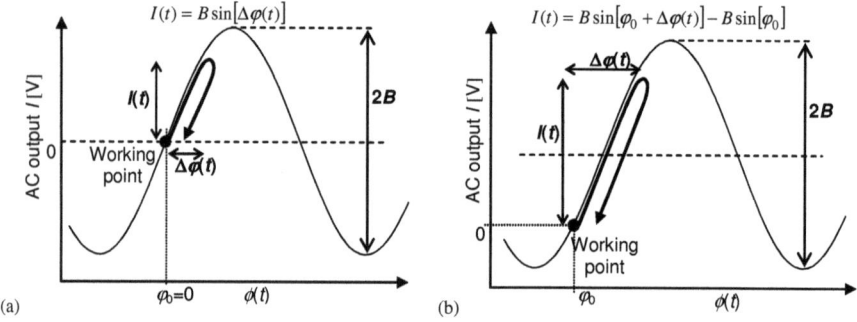

Fig. 3.4. Phase shift measurement. (a) The working point is in the zero-crossing point of the sinusoid, (b) the working point is not in the zero-crossing point of sinusoid. In case (b) high sensitivity in longer range of phase $\Delta\varphi$ is reached. Since the AC output cut-off frequency is 30 kHz and the stress repetition rate is 1 Hz, the signal in the working point is always zero volts.

3 Dual-beam interferometer

The detector has two separated outputs: DC and AC output (with cut-off frequency 30 kHz). The heat-induced phase shift $\Delta\varphi(t)$ is measured by the AC output. In the AC output the offset $A = B\sin(\varphi_0)$ in Eq. 3.3 is filtered out. Due to this the $I(t)$ is always zero in the working point φ_0 and a constant $B\sin(\varphi_0)$ has to be subtracted from $I(t)$ in Eq. 3.3 (see Fig. 3.4b). The signal in the AC channel is then:

$$I(t) = B\sin[\varphi_0 + \Delta\varphi(t)] - B\sin[\varphi_0]. \tag{3.4}$$

The phase shift is then calculated from the measured signal $I(t)$ by following equation:

$$\Delta\varphi(t) = \arcsin\left(\frac{I(t) + B\sin(\varphi_0)}{B}\right) - \varphi_0. \tag{3.5}$$

If the working point $\varphi_0 = 0$ rad, the Eq. 3.5 simplifies to $\Delta\varphi = \arcsin[I(t)/B]$. If the term in the parenthesis reaches value ± 1 (signal $I(t)$ reaches a local extreme of the sinusoid), the resulting phase is modulo π. For small phase shift the unwrapping is not needed.

<u>Determination of constant B</u>

The AC and DC outputs of the detector have a different amplification. The ratio of the AC and DC output amplification was experimentally found to be around 0.1 (exact value depends on the particular detector). The constant B_{DC} for the DC output in a particular point (x, y) can be measured e.g. by introducing the mechanical vibrations to the reference mirror. The corresponding constant B for the AC output is then calculated from the amplification ratio $B = 0.1 * B_{DC}$.

3.3.1 Signal timing

In the nanosecond time scale the speed of the electrical signal in the cables and the speed of the light in the air is important to take into account. The signals recorded by the oscilloscope have to be correctly shifted in time by the delays caused by the cables and by the distances of the instruments.

3 Dual-beam interferometer

The timing diagram of the dual-beam interferometer for the case of TLP and vf-TLP stressing is shown in Fig. 3.5. From the delays A, B, C, and D the correct delay of the voltage (current) waveform and the phase waveform is calculated. In the case of TLP the current probe is used to measure the current through the DUT, in the case of vf-TLP the voltage on the DUT is calculated from the difference of the incident waveform and the waveform reflected from the DUT.

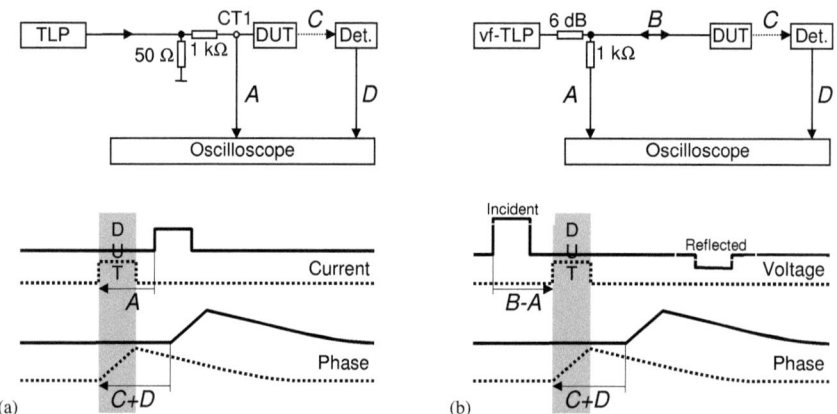

Fig. 3.5. Timing diagram of the dual-beam interferometer for (a) TLP and (b) vf-TLP stressing. Bold lines indicate the signal on the oscilloscope, dashed lines and grey rectangles the signal on the DUT.

The cables used in the setup are coaxial 50 Ω matched cables. An electrical signal passes one meter of the cable in time 4.8-5.1 ns (depending on the cable propagation coefficient), the light passes one meter of the air in time 3.3 ns. Therefore to keep the setup precision at 0.4 ns, the electrical path has to be measured with precision better than 8 cm and the optical path with precision better than 12 cm.

3.4 Error sources

In the following part the influence of the vibrations, optical feedback and electromagnetic pick-ups on the measurement is analysed.

3.4.1 Vibrations

The Michelson interferometer is generally very sensitive to any mechanical vibrations of the setup components. This limits the sensitivity of the phase measurement. In the measurement of phase changes in the nanosecond time range the mechanical vibrations can be neglected, since they are much slower. They have influence on the balancing of the working point only. Therefore in the case of repetitive measurement where the waveforms are averaged, the vibrations have to be minimised or compensated by a feedback to the reference mirror. In our setup the reference mirror is moved with piezo crystal to balance the working point. The correct voltage on the crystal is set according to the signal from the DC detector channel.

3.4.2 Optical feedback

Part of the laser light is reflected from the optical components and returned back to the laser diode. This is called optical feedback. The feedback can cause amplitude fluctuation, frequency shift, modulation bandwidth limitation, noise and even a damage. With a simple experimental setup shown in Fig. 3.6a the relevance of the feedback was estimated. In Fig. 3.6c a special case of the signal from the DC and AC detector output is shown, where the feedback is unstable in time. The diode is switching between two operational modes. The standard deviation of noise in the AC channel is either 2.7 mV or 14 mV. Simultaneously the changes in the power of the laser beam are detected in the DC channel.

The noise is transmitted to the phase according to Eq. 2.3. The noise standard deviation may reach values up to 45 mV in some cases. If in such case the interference signal amplitude is 50 mV, then the phase error is around 0.7 rad (calculated by Eq. 2.3).

To remove the optical feedback a Faraday isolator is placed behind the laser diode, see Fig. 3.6b. The isolator transmits the light in one direction and prevents the reflected or scattered light to return back to the laser. The isolation is 46 dB (produced by Döhrer Elektrooptik) and it significantly reduces the noises on the AC output allowing thus single pulse measurement with precision better than 0.1 rad.

3 Dual-beam interferometer

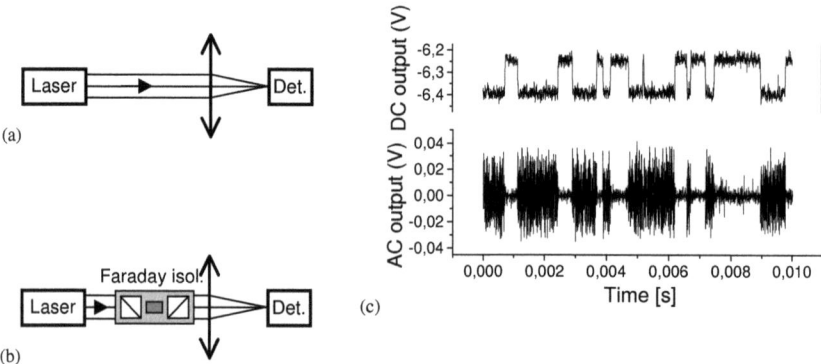

Fig. 3.6. (a, b) Experimental setups for the optical feedback measurement, (c) example of the feedback induced noise.

3.4.3 Electromagnetic pick-ups

The pulser, connectors, bonding wires and sample itself emit the electromagnetic radiation. Since the stress pulses in the setup have amplitude up to 10 A, this radiation is received by the laser diode and its power source and disturbs the laser generation. Consequently wide spectral range electromagnetic pick-ups appear in the phase signal during and after the electrical pulse, see Fig. 3.7. The amplitude of the pick-ups depends on the current amplitude, setup and cable geometry, cable lengths and DUT too. The pick-ups limit strongly the phase sensitivity and time resolution. The same pick-ups are reproduced from pulse to pulse and therefore can not be reduced by averaging. Spectral filtering of the pick-ups would decrease the time resolution of the setup.

To decrease the pick-ups the metallic shielding boxes for the DUT and for the laser diodes are used, see Fig. 3.8. The laser diodes are supplied from the batteries and placed in a metallic box isolated from the optical table. The alternating electromagnetic field induces an alternating current in the metal, which exponentially decreases with the depth. The penetration depth δ is given by relation [Mei92]:

$$\delta[\mu m] = 64\eta / \sqrt{f[MHz]} \qquad (3.6)$$

3 Dual-beam interferometer

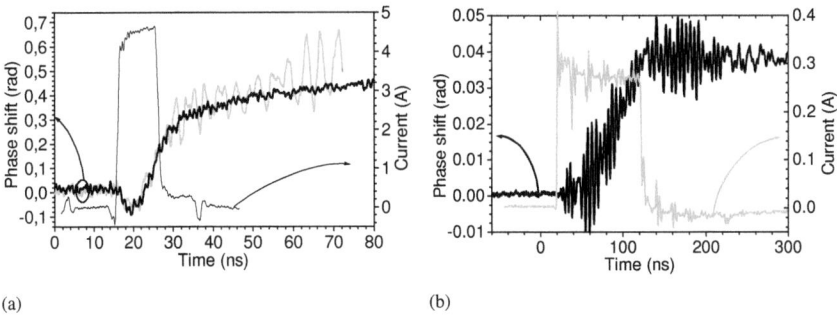

(a)   (b)

Fig. 3.7. Examples of the electromagnetic pick-ups in the phase for 10 ns and 100 ns current pulse. In case (a) the black line shows the signal with suppressed pick-ups, recorded in the shielded setup.

where η is a material factor and f is the electrical field frequency. δ characterises the depth where the alternating current density decreases to 1/e. Some examples of δ for various materials and field frequencies are given in Table 3.2. For shielding of the DUT the Cu box of wall thickness 1 mm is used, for shielding of the laser diodes the Al box of wall thickness 2.5 mm is used. The effect of shielding is shown in Fig. 3.7a, where the black line shows the signal with suppressed pick-ups.

Fig. 3.8. Shielding against the pick-ups. (a) Laser and the battery power supply are electrically isolated from all other instruments, (b) covered box for laser shielding, (c) shielding of the DUT.

3 Dual-beam interferometer

Table 3.2. The penetration depth of an alternating current.

Material	η	δ @ 10 GHz	δ @ 1 GHz	δ @ 100 MHz	δ @ 10 MHz
Steel	6.7	4.3 µm	14 µm	43 µm	140 µm
Al	1.35	0.86 µm	2.7 µm	8.6 µm	27 µm
Cu	1.03	0.66 µm	2.1 µm	6.6 µm	21 µm

3.5 Examples

3.5.1 Example 1 – Phase calibration

To compare the phase measurement precision of the dual-beam interferometer with the heterodyne scanning setup a simple diode structure was measured in both setups and the result was compared. The IR image of the diode is shown in Fig. 3.9a. One example for the reverse biased diode and 250 mA current stress pulse is shown in Fig. 3.9b. The phase shifts measured in both setups overlap very well.

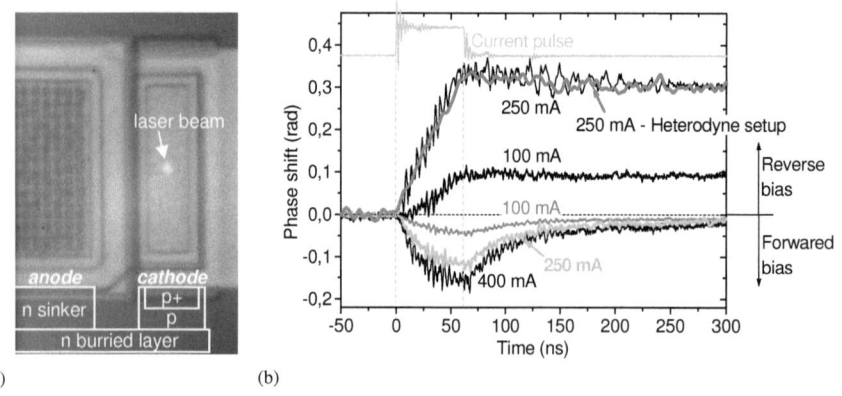

Fig. 3.9. (a) IR image of a diode. (b) Phase shift in the diode structure for forward (negative phase shift) and reverse (positive phase shift) bias under the current stress pulses of 60 ns duration with different amplitude. The signals are five times averaged. For comparison the signal from the heterodyne setup is also shown.

3 Dual-beam interferometer

The thermal and the free carrier effects can be well distinguished in the setup. In the reverse biased diode the thermal effect (positive values) is dominant, in the forward biased diode the free carrier effect (negative values) is dominant [Byc02, Pog02c], see Fig. 3.9. During the stress pulse, the thermal signal increases approx. linearly, but the free carrier signal saturates. After the pulse end the thermal signal decreases much slower than the free carrier signal.

3.5.2 Example 2 – Separation of thermal and free carrier contribution

In some cases the thermal and the free carrier contribution to the phase shift are both relevant and present simultaneously at the same position. The separation of these two signals is difficult [Gol03, Tha98]. An approximate estimation of these two contributions can be done, as will be shown on an example bellow.

The npn transistor ESD protection device with a lateral and buried layer collector and a short-circuited base/emitter was investigated [Bla02, Bla03, Dub03]. In order to find out if both the lateral npn and vertical npn transistors are active simultaneously, we have focused the two laser beams in the corresponding positions on the device, see Fig. 3.10a. Fig. 3.10b shows the phase shift evolution at these two positions during a 3 A and 10 ns long vf-TLP pulse.

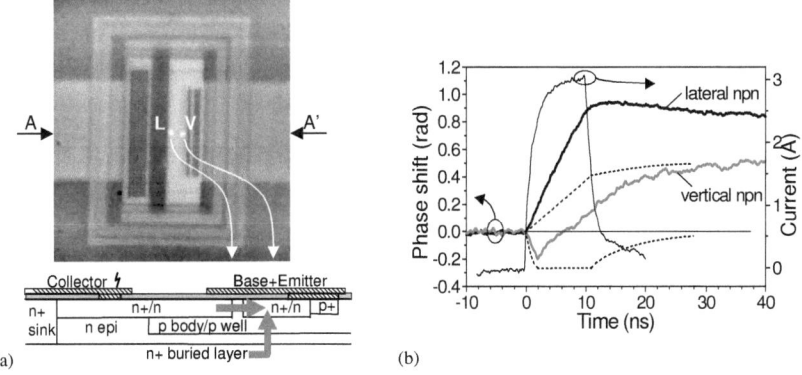

Fig. 3.10. (a) Backside IR image of the npn structure (normal bipolar) with marked laser beam positions and device cross section, (b) the phase shift development in marked positions from (a).

At the position of the lateral npn transistor, immediately with the beginning of the pulse the phase shift increases nearly linearly due to the self-heating effect in the impact ionisation region of the base-collector junction. After the end of the pulse the heat spreads to the surrounding. At the position of the vertical transistor a rapid decrease occurs immediately after the beginning of the pulse, which is followed by a phase increase. The start signals from the lateral and vertical npn is well aligned, indicating that there is trigger delay less than 0.4 ns.

The signal at the vertical npn is caused by a superposition of a free carrier negative signal, dominant in a short time scale and a thermal signal, which is much slower (Fig. 3.10b). These two components are tentatively separated and plotted, see dotted lines. The steady state of the free carrier signal is obtained after some 2 ns. Since this moment the free carrier signal is constant and the increasing of the phase is caused by the thermal signal only. Assuming this the thermal signal during the pulse can be extracted. By subtracting the thermal signal and the measured phase (total signal) the free carrier signal is determined.

After the pulse end the free carrier signal decreases in range of several nanoseconds to tens on nanoseconds (this depends on the device) and the thermal signal starts to be dominant (thermal signal decreases in the range of microseconds). By extrapolation of this thermal phase component toward the stress pulse the thermal component just after the stress pulse end is obtained. The thermal component exhibits a small rise after the pulse end in this particular case, which is due to the heat transfer from the surrounding region.

3.5.3 Example 3 – Measurement of small trigger delay

The moving current filament in a coupled npn/pnp ESD protection device was investigated with the 2D TIM setup in Chapter 2.8.2. At higher stress currents the lateral transistor (see Fig. 2.50f) is triggered. The delay between triggering of the vertical npn and lateral pnp transistor was investigated by the dual-beam interferometer [Byc02].

The IR image of the device with the two laser spots is shown in Fig. 3.11a. The device was stressed using the vf-TLP technique with 20 ns pulse duration. The current through the device was 2, 3.5 and 5 A. The laser beam marked by L is placed over the active region of the lateral transistor and the laser beam marked by V is placed over the active region of the vertical transistor. Fig. 3.11b shows the time evolution of the phase shift in these two positions. In the

3 Dual-beam interferometer

area 'I' the current flows homogeneously in the whole device. The vertical transistor is triggered 3 ns after the current pulse beginning, see positive phase shift in area 'II'. After next 2 ns the lateral transistor triggers, see a gentle negative phase shift in area 'III'. The steady state of the free carrier signal is obtained after next 2 ns and then the phase shift starts to increase due to the thermal effect. We also notice, that after the lateral transistor is triggered, the current is redistributed between these two transistors. Since the total current is constant, the slope of the phase shift at vertical transistor decreases, compare the grey lines 'A' and 'B'.

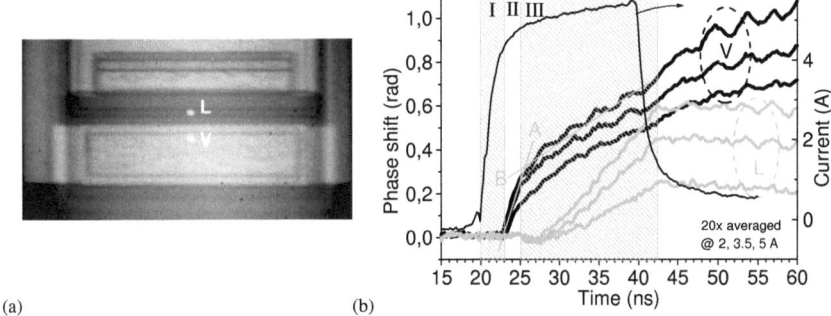

Fig. 3.11. The coupled npn/pnp ESD protection device. (a) The IR image of the device with marked laser beams spots. (b) The phase shift measured at position of lateral (L) and vertical (V) transistor under different stress conditions: 2, 3.5 and 5 A.

3.5.4 Example 4 – Measurement of unrepeatable phenomena

The pulse to pulse instability of the vertical DMOS transistor was investigated by the 2D TIM setup in Chapter 2.8.3. It was shown that the current filament creation is a random process and that the filaments do not spread, but move from cell to cell. From the 2D measurements it is still not clear if the current filaments coexist at the same time, or if the current oscillates between the hot spots within the period lower than 5 ns (time resolution of the 2D TIM setup) [Den04]. To answer this question, the DMOS was investigated in the dual-beam interferometer.

The two laser beams were focused near the drain contact, see Fig. 3.12a. Three typical examples of the phase shift evolution are shown in Figs. 3.12b-d. The three graphs show three

3 Dual-beam interferometer

triggering modes: in case (b) single or more filaments were created and they appeared in the measured positions at different time instants, in case (c) only one filament appeared in position 2 and in case (d) two coexisting filaments are shown. The last case confirms the coexistence of two current filaments. After some time the phase increasing stops and the phase stays constant, indicating that the filament has moved to another region. The phase waveforms differ from pulse to pulse in the risetime (different instantaneous P_{2D} in the current position) and in the amplitude (different dissipated energy in the local position). In Fig. 3.12d the phase shift in position 2 starts to increase again at time 90 ns indicating that the current filament has returned to this position or one of the existing filaments proceeded to this area.

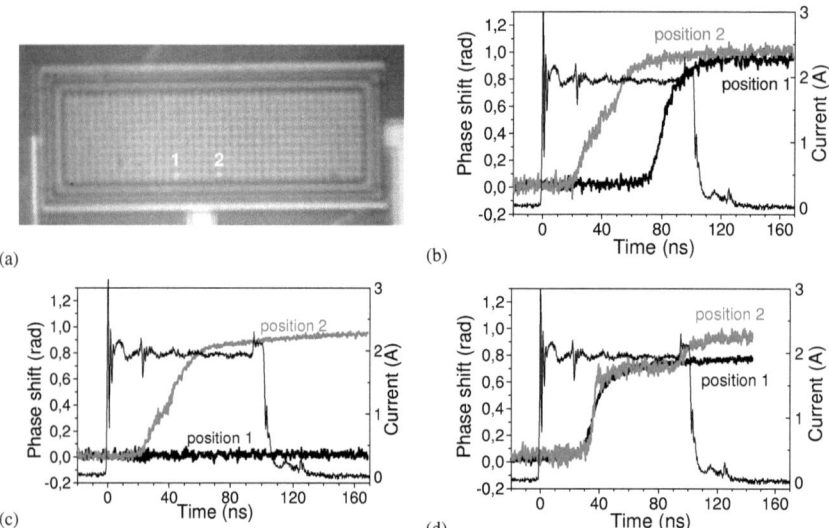

Fig. 3.12. (a) IR image of the vertical DMOS with 440 cells. The laser beam positions are marked. (b-d) Examples of the phase shift: (b) first current filament is created in position 2, later next one in position 1, (c) only one filament is created in position 2, (d) two coexisting filaments.

3.6 Summary

3 Dual-beam interferometer

The dual-beam setup based on Michelson interferometer has been introduced. The setup time resolution is 0.4 ns and spatial resolution 2 µm. The phase sensitivity of the setup depends on the DUT reflectivity, for the metal surface the phase bellow 0.1 rad can be detected. For devices with repeatable internal behaviour the sensitivity can be improved by averaging directly in the oscilloscope down to milliradians. To reach this sensitivity, the optical feedback had to be avoided and the electromagnetic pick-ups suppressed. In this setup the phase shift at two positions in maximal relative distance 230 µm is measured simultaneously. This allows measurement of sub-nanosecond trigger delays between these two points.

4 Summary

Two optical techniques based on the backside laser interferometry for the characterisation of semiconductor devices under a short electrical pulse were developed and demonstrated. These non-destructive techniques provide information about the thermal and current flow distributions in semiconductor devices during a single stress pulse, which can not be obtained by any other method.

The 2D TIM setup was developed and built in two variants: the laboratory variant, flexible for research purposes and the compact probe station variant, suitable for implementation in the industry. With the 2D TIM technique the thermal distribution in the device at two time instants can be obtained during a single stress pulse with a time resolution of 5 ns, spatial resolution up to 2 µm and variable field of view from 230 µm to 6 mm. The two time instants can be chosen arbitrarily up to several seconds.

The method based on *Fourier transform* was found to be the optimal method for the data analysis. The phase extraction methodology was adapted in order to optimise the evaluation of interferograms recorded in the 2D TIM setup. It was shown that the low-frequency undulations in the extracted phase originate from the device topology and that they are a consequence of overlapping of the Fourier spectrum components and the noise filtering. Therefore an optimised adaptive spectrum filter was proposed, which reduces the undulations and simultaneously minimises the number of the phase defects and noise in the extracted phase.

Since the unwrapping of the wrapped phase is very sensitive to the phase defects and noise, several phase unwrapping methods were tested. Based on this testing a phase pre-processing method was proposed, which rapidly decreases the probability of a phase step error creation and error propagation to the surrounding area, isolating thus the phase artifacts. Pre-processing reduces the necessity of phase unwrapping of the whole phase map and significantly speeds up the phase unwrapping process.

The device properties like backside polishing, geometry, topology and surface reflectivity were classified and their effect on the phase shift was analysed. The best reflectivity has the power metal where the phase extraction precision is 0.1 rad and the worst have the contact vias where the phase sensitivity degrades up to 0.8 rad due to low S/N ratio. The device edges have marginal influence on the extracted phase shift since the phase errors are localised around the

4 Summary

device edges. In the case of laser with coherence length 2 mm and device substrate thickness 400 µm the phase error resulting from the multiple interference within the substrate is maximally 0.1 rad and for thinner substrates it increases. The effect of the laser pulse duration on the phase measurement was found to be negligible. It was observed that the laser pulse-to-pulse instability brings additional phase variation up to 0.2 rad. Form this we conclude that the typical phase sensitivity of the 2D measurement is around 0.5 rad and it depends mainly on the device topology.

It was shown that by measuring of the phase shift at two nearby time instants and by using an approximation for the time derivative the instantaneous 2D power dissipation density can be extracted. If the size of the device is smaller than the thermal diffusion length, the experimentally obtained 2D power dissipation density agrees with expected value and shape, otherwise its amplitude is overestimated.

Using the 2D TIM setup the moving current filaments in the ESD protection element has been detected. In the DMOS transistor structure inhomogeneous current flow dynamics was revealed for the first time and the measurement of the current dynamics during a destructive stress pulse was possible to perform.

Thanks to the laser wavelength tuning the optimal wavelength for the temperature-induced absorption changes measurement was established. The absorption measurement technique was found to show smaller dynamic range and sensitivity than the interferometric technique. Both techniques are suitable for measurements in millisecond time range too.

The dual-beam interferometer enables independent measurement in two positions on the device in maximal distance 230 µm with 0.4 nanosecond time resolution, which is especially convenient for the investigation in CDM time domain. The spatial resolution is approx. 2 µm and the phase sensitivity is better than 0.1 rad for the single pulse measurement, depending on the reflectivity of the device. For devices without pulse to pulse instability in current flow the signal can be averaged during several pulses to improve the S/N ratio. The sensitivity is limited by the electromagnetic pick-ups, which originate from the high power short stress pulses. The pick-ups were reduced by an appropriate shielding of the setup. With this setup the sub-nanosecond trigger delays and current flow instabilities during a single stress pulse were measured, as it was shown in examples, which makes this setup an unique tool in the device analysis.

The results of current flow distribution and triggering behaviour were used for verification of simulated data (3D simulation models) of our industrial partners.

Appendix

Analysis software for phase shift extraction

Processing overview

The simplified data flow chart of the interferogram analysis software is shown in Fig. A.1. The input files are the interferogram bitmaps, which were saved during the experiment. The flow chart contains two branches: manual processing mode and automatic processing mode. In the manual mode the interferogram processing is fully controlled by the user whereas in the automatic mode the processing is fully automated. Notice also that in the automatic mode the multiple interferograms can be processed at once. The result of the processing is the 2D phase map that can be further manually refined before saving. The output of the program consists of the data files such as 2D phase maps and their cross-sections that can be saved on hard disc.

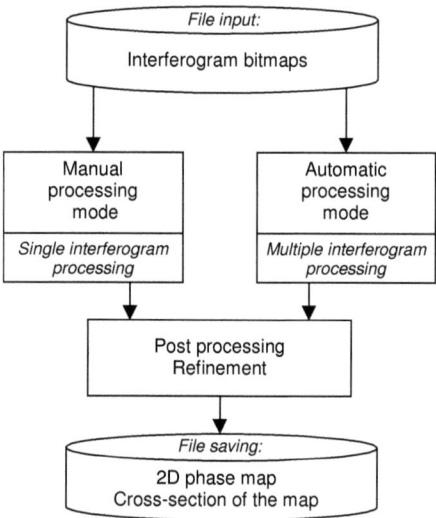

Fig. A.1. Simplified data flow chart of the interferogram analysis software.

Appendix

Manual processing mode

A detailed flow chart of the interferogram processing is shown in Fig. 2.18. The white arrows in Fig. 2.18 mark the processing flow for the manual mode. The flow chart of the data processing consists of 9 steps described in Chapter 2.5.1. More details about these steps in the manual processing mode are described bellow:

1) **Input interferograms.** User loads a single interferogram bitmap file of the unstressed device from the file exchange interface.
2) **FFT.** The spectrum of the loaded interferogram is calculated with the Fast Fourier transform algorithm and shown to the user. As well the user can visualise the absolute and phase values, real and imaginary parts of the spectrum in linear or logarithmic scales.
3) **Spectrum filtering**. In the manual mode the spectrum filtering can be done in four different ways:
   a) automatic filtering by an optimal filter described in Chapter 2.5.2.3,
   b) load the filter from the hard disk and apply it to the spectrum,
   c) using an interactive graphic interface the user can create its own spectrum filter and apply it to the spectrum. The special helping functions such as filter symmetrisation were implemented. The created spectrum filter can be saved in a file for further exploitation,
   d) combination of two previous methods: load existing spectrum filter, modify it with the graphical interface and apply it to the spectrum.

Optionally the filtered spectrum can be shifted to the point [0, 0] of the coordinate system (spatial heterodyning). For some types of interferograms this operation gives advantages in the unwrapping process.

4) **Inverse FFT.** User can visualise the absolute value, real and imaginary parts of the calculated result on the screen. This information helps to investigate the nature of artifacts present in the 2D phase map.
5) **Phase mod 2π.** The wrapped phase is calculated from the real and imaginary parts. Result is shown in a separate window. At this point user is asked to input of the filename of the interferogram of the device under the stress. The stressed interferogram will be loaded and processed (steps 1-5) with exactly the same parameters as it was done with the reference interferogram.

Appendix

6) **Subtraction.** The wrapped phase maps (reference and stressed) are subtracted.
7) **Pre-processing.** This was described in Chapter 2.5.3.1.
8) **Unwrapping.** The several unwrapping algorithms were implemented in the software due to different reliability and speed:
   a) *Straightforward* **algorithm** *(fast but not reliable, causes spreading of artifacts)*. The unwrapping procedure processes the columns or rows of the 2D phase map. User can choose row processing, column processing and direction in which the processing will be done. For instance for the column processing the user can chose from top to down and from down to top directions.
   b) *Spiral* **algorithm** *(slower than (a) algorithm and more reliable)*. The unwrapping process is done along a spiral path. The neighbour points to the processing point are taken into account during unwrapping.
   c) *Pixel-queue* **algorithm** *(slower than (b) algorithm but reliable)* is similar to the spiral algorithm but more accurately takes into account the neighbour points.
   d) *Minimum spanning tree* **algorithm** *(slowest but most reliable)* Points where unwrapping is more stabile are processed first. Unwrapping process automatically choose the optimal path for the phase unwrapping according the priorities.
9) **Post processing and result saving** – several algorithms for data visualisation and processing are implemented as well as additional phase unwrapping procedures are available:
   - data filtering with median filter or Savitzky-Golay filter for reduction of local artifacts
   - invert operation for 2D phase map for correction of phase shift sign
   - subtraction of the plane for removing of the phase surface slope
   - add constant for 2D phase map for removing artificial constant background
   - line and circular cross section of the 2D phase map for detailed visualisation of local regions and comparison of several images
   - cropping of the 2D phase map for removing of useless margins
   - zooming In/Out of phase map
   - adding, subtracting, multiplying and dividing of the phase by a different phase map
   - visualisation of the 2D map in different scales and different colours
   - calculation of space derivatives of the phase shift using the Savitzky-Golay algorithm

   The post processing suppresses the artifacts, which are mainly related to areas of low reflectivity of the sample and pulse to pulse laser instability.

Appendix

The result can be saved in different file formats: binary files, picture file (user could choose following formats: *.bmp, *.tif, *.jpg), text file (ASCII data). The cross section data can be saved as text file (ASCII data file) for further analysis and plotting in programs like Microcal^{TM} Origin^{TM} or Microsoft® Excel.

Notice that in the manual mode user can save or load result (or previously saved data) of the data processing at any step (1-9). It is also possible to make a cross sections of all data at any step.

Automatic processing mode

This mode does processing (see steps 2-7 in Fig. 2.18) of the multiple interferograms at once fully automatically. The algorithm uses the same processing core, which was described for manual processing mode but with automatic masking of the spectrum.

The file-loading interface was enhanced in comparison with manual mode (see Fig. A.2). Following file interface features were implemented:
- browsing through the all interferogram images in the chosen directory
- previewing of the selected interferogram file on the screen
- choice of the multiple interferogram files for processing
- automatic saving of the results for all processed interferograms
- cross section for all interferograms in the same area at once
- colour map and range changes of all resulting phase maps at once

The screenshot of the analysis software, demonstrating the above features, is shown in Fig. A.3.

Appendix

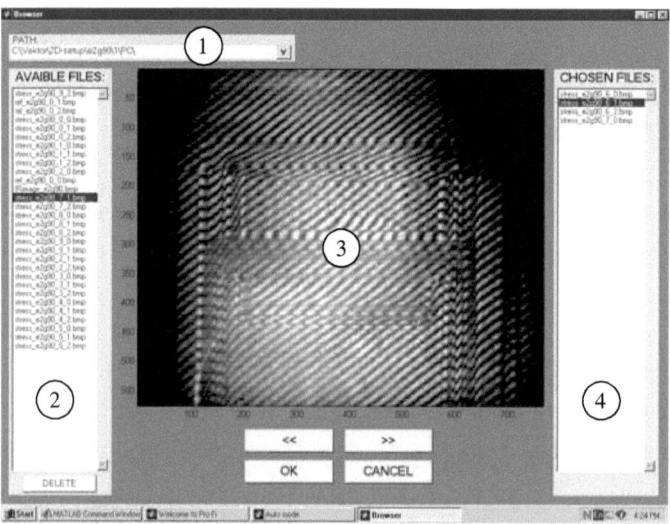

Fig. A.2. The file loading interface for the auto-mode. The list of the files of the directory (1) is shown in (2) and the selected file is viewed in (3). The files chosen for the evaluation are in (4).

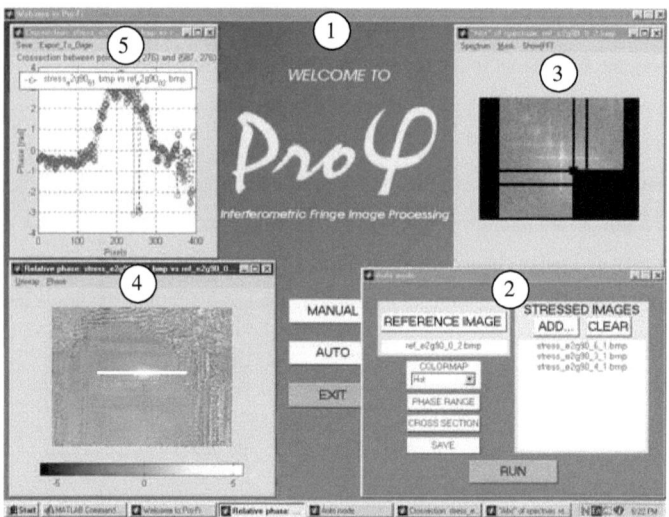

Fig. A.3. The screenshot of the fringe analysis software *Proφ*. (1) - main window, (2) - operational window of auto mode, (3) - example of a spectrum, (4) - example of a phase map and (5) - cross section of a phase map on the marked position (white line in 4).

Bibliography

[Aim00] G. Aiming, C. Lei, C. Jinbang and Z. Rihong. Research on processing technique for interference pattern with step using phase-shift interferometry. *Adv. Opt. Man. and Test. Techn., Proc. of SPIE*, vol. 4231, 2000, pp. 371-374.

[Ame92] A. Amerasekera and J. Verwey. ESD in integrated circuits. *Qual. Rel. Eng. Int.*, vol.8, 1992, pp. 259-272.

[Ame93] A. Amerasekera, M.-C. Chang, J. A. Seitchik, A. Chatterjee, K. Mayaram and J.-H. Chern. Self-heating effects in basic semiconductor structures. *IEEE Trans. Electron Devices*, vol. 40, 1993, pp. 1836-1844.

[Ame95] A. Amerasekera and C. Duvvury. ESD in silicon integrated circuits. Wiley, Chichester, 1995.

[Bar95] D. L. Barton and P. Tangyunyong. Fluorescent microthermal imaging - theory and methodology for achieving high thermal resolution images. *Microelectr. Eng.*, vol. 31, no. 1-4, 1995, pp. 271-279.

[Ber90] M. Bertolotti, V. Bogdanov, A. Ferrari, A. Jascow, N. Nazorova, A. Pikhtin, and L. Schirone. Temperature dependence of the refractive index in semiconductors. *J. Opt. Soc. Am. B*, vol. 7, 1990, pp. 918-922.

[Bla87] A. Black, C. Courville, G. Schultheis and H. Heinrich. Optical sampling of GHz charge density modulation in silicon bipolar junction transistors. *Electron. Lett.*, vol. 23, no. 15, 1987, pp. 783-784.

[Bla02] M. Blaho, D. Pogany, L. Zullino, A. Andreini and E. Gornik. Experimental and simulation analysis of a BCD ESD protection element under the DC and TLP stress conditions. *Microel. Reliab.*, vol. 42, 2002, pp. 1281-1286.

[Bla03] M. Blaho, D. Pogany, E. Gornik, L. Zullino, E. Morena, R. Stella, A. Andreini, H. Wolf and H. Gieser. Internal behavior of BCD ESD protection devices under very-fast TLP stress. *Proc. of 41st Annual International Reliability Physics Symposium*, Dallas, Texas, 2003, pp. 235-240.

[Ble92] H. Bleichner, M. Bakowsky, M. Rosling, E. Nordlander, J. Vobecky, M. Lundquist and S. Berg. A Study of turn-off limitations and failure mechanisms of GTO-thyristors

Bibliography

by means of 2-D time-resolved optical measurements. *Solid State Electronics*, vol. 35, 1992, pp. 1683-1695.

[Bon86]   D. J. Bone, H.-A. Bachor and R. J. Sandeman. Fringe-pattern analysis using a 2D Fourier transform. *Appl. Opt.*, vol. 25, 1986, pp. 1653-1660.

[Bon91]   D. J. Bone. Fourier fringe analysis: the two-dimensional phase unwrapping problem. *Appl. Opt.*, vol. 30, 1991, pp. 3627-3632.

[Bor80]   M. Born and E. Wolf. Principles of Optics. 6th Ed., Pergamon Press, Oxford, 1980.

[Bre97]   G. Breglio, S. Pica, P. Spirito and A. Tartaglia. Experimental detection of time dependent temperature maps in power bipolar transistors. *Proc. THERMINIC'97*, Cannes, France, 1997, pp. 127-132.

[Bre86]   B. Breuckmann and W. Thieme. New computer-aided system for automatic fringe analysis in optical metrologies. *SPIE Proc. Autom. Opt. Inspec.*, vol. 654, 1986, pp. 237-241.

[Byc02]   S. Bychikhin, V. Dubec, M. Litzenberger, D. Pogany, E. Gornik, G. Groos, K. Esmark, M. Stecher, W. Stadler, H. Gieser and H. Wolf. Investigation of ESD protection elements under high current stress in CDM-like time domain using backside laser interferometry. *J. of Electrostatic*, vol. 59, 2003, pp. 241-255.

[Chi92]   N. H. Ching, D. Rosenfeld and M. Braun. Two-dimensional phase unwrapping using a minimum spanning tree algorithm. *IEEE Tran. on Image Proc.*, vol. 1, 1992, pp. 355-365.

[Cse95]   A. Csendes, V. Szekely and M. Rencz. Thermal mapping with liquid crystal method. *Elec. and Opt. Beam Testing of Electr. Dev.*, Wuppertal (Germany), 1995.

[Deb93]   G. Deboy and J. Kölzer. Fundamentals of light emission from silicon devices. *Semicond. Sci. Technol.*, vol. 9, 1993, pp. 1017-1031.

[Den03]   M. Denison, M. Blaho, D. Silber, J. Joost, N. Jensen, M. Stecher, V. Dubec, D. Pogany and E. Gornik. Hot spot dynamics in quasi vertical DMOS under ESD stress. *Proc. ISPSD'03*, 2003, pp. 80-83.

[Den04]   M. Denison, M. Blaho, P. Rodin, V. Dubec, D. Pogany, D. Silber, E. Gornik and M. Stecher. Moving current filaments in integrated DMOS transistors under short-duration current stress. *IEEE Trans. Electron Dev.*, vol. 51, 2004, pp. 1695-1703.

Bibliography

[Dil97] S. Dilhaire, T. Phan, E. Schaub and W. Claeys. High sensitivity and high resolution differential interferometer: Micrometric polariscope for thermomechanical studies in microelectronics. *Microelectr. Reliab.*, vol. 37, no. 10-11, 1997, pp. 1587-1590.

[Dil99] S. Dilhaire, J. Altet, S. Jorez, E. Schaub, A. Rubio and W. Claeys. Fault localisation in ICs by goniometric laser probing of thermal induced surface waves. *Microelectr. Reliab.*, vol. 39, 1999, pp. 919-923.

[Don90] V. M. Donnelly and J. A. McCaulley. Infrared-laser interferometric thermometry: a non-intrusive technique for measuring semiconductor wafer temperatures. *J.Vac.Sci.Technol.*, vol. A8, no. 1, 1990, pp. 84-92.

[Dub03] V. Dubec, S. Bychikhin, M. Blaho, D. Pogany, E. Gornik, J. Willemen, N. Qu, W. Wilkening, L. Zullino and A. Andreini. A dual-beam Michelson interferometer for investigation of trigger dynamics in ESD protection devices under very fast TLP stress. *Microel. Reliab.*, vol. 43, 2003, pp. 1557-1561.

[Dub04b] V. Dubec, S. Bychikhin, M. Blaho, M. Heer, D. Pogany, M. Denison, N. Jensen, M. Stecher, G. Groos and E. Gornik. Multiple-time-instant 2D thermal mapping during a single ESD event. *Microel. Reliab.*, vol. 44, 2004, pp. 1793-1798.

[Dub04a] V. Dubec, S. Bychikhin, D. Pogany, E. Gornik, G. Groos and M. Stecher. Error analysis in phase extraction in a 2D holographic imaging of semiconductor devices. *Proc. SPIE, Practical Holography XVIII: Materials and Applications*, vol. 5290, 2004, pp. 233-242.

[Dwy90] V. M. Dwyer, A. J. Franklin and D. S. Campbell. Thermal failure in semiconductor devices. *Solid-St. Electron.*, vol. 33, 1990, pp. 553-560.

[EIA/JEDEC] EIA/JEDEC. Standard Test Method A114-A Electrostatic Discharge (ESD) Sensitivity Testing Human Body Model (HBM), EIA/JESD22-C114-A, 1997.

[ESD-R] Electrostatic discharge association, Rome, NY, USA. ESD STM5.1-2001 Electrostatic Discharge Sensitivity Testing – Human Body Model (HBM) Component Level, 2001.

[Esm00] K. Esmark, C. Fürböck, H. Gossner, G. Groos, M. Litzenberger, D. Pogany, R. Zelsacher, M. Stecher and E. Gornik. Simulation and experimental study of temperature distribution during ESD stress in smart-power technology ESD protection devices. *Proc. IRPS'2000*, San Jose, 2000, pp. 304-309.

[Fer68] J. L. Fergason. Liquid crystals in nondestructive testing. *Appl. Opt.*, vol. 7, no. 9, 1968, pp. 1729-1737.

[Fri90]   J. Fritz and R. Lackmann. Optical beam induced currents in MOS transistors. *Microel. Eng.*, vol. 12, 1990, pp. 381-388.

[Fur99]   C. Fürböck, R. Thalhammer, M. Litzenberger, N. Seliger, D. Pogany, E. Gornik and G. Wachutka. A differential backside laser probing technique for the investigation of the lateral temperature distribution in power devices. *Proc. ISPSD'99*, Toronto, 1999, pp. 193-196.

[Gie99]   H. Gieser. Verfahren zur Charakterisierung von integrierten Schaltungen mit sehr schnellen Hochstrompulsen. Shaker Verlag, Aachen, FRG, 1999.

[Gie98]   H. Gieser and M. Haunschild. Very fast transmission line pulsing of integrated structures and the charged device model. *IEEE Trans. Comp. Pack. Manuf. Tech. - Part C*, vol. 21, 1998, pp. 278-285.

[Gla96]   J.-Y. Glancet and S. Berne. A practical system for hot spot detection using fuorescent microthermal imaging. *Microelectr. Reliab.*, vol. 36, 1996, no. 11/12, pp. 1811-1814.

[Gol93]   M. Goldstein, G. Sölkner, and E. Gornik. Heterodyn interferometer for the detection of electric and thermal signals in integrated circuits through the substrate. *Rev. Sci. Instr.*, vol. 64, 1993, pp. 3009-3013.

[Hee05]   M. Heer, V. Dubec, M. Blaho, S. Bychikhin, D. Pogany and E. Gornik. Automated setup for thermal imaging and electrical degradation study of power DMOS devices," accepted for ESREF'05.

[Hei86b] H. K. Heinrich, D. M. Bloom and B. R. Hemenway. Noninvasive sheet charge density probe for integrated silicon devices. *Appl. Phys. Lett.*, vol. 48, 1986, pp. 1066-1068.

[Hei86a] H. K. Heinrich, B. R. Hemenway, K. A. McGroddy and D. M. Bloom. Measurements of real-time digital signals in a silicon bipolar junction transistor using a noninvasive optical probe. *IEEE Electron Device Lett.*, vol. 22, 1986, pp. 650-652.

[Her95a] I. P. Herman. Real-time optical thermometry during semiconductor processing. *IEEE J. Sel. Top. Quant. Electr.*, vol. 170, no. 413, 1995, pp. 1047-1053.

[Her94]   P. Herve and L. K. Vandamme. General relation between refractive index and energy gap in semiconductors. *Infrared Phys. Technol.*, vol. 35, no. 4, 1994, pp. 609-615.

[Her98]   C. Herzum, C. Boit, J. Kölzer, J. Otto and R. Weiland. High resolution temperature mapping of microelectronic structures using quantitative florescence microthermography. *Microelectr. Journal*, vol. 29, no. 4-5, 1998, pp. 163-170.

[Hua02] M. J. Huang and C.-J. Lai. Phase unwrapping based on a parallel noise-immune algorithm. *Opt. &Laser Tech.*, vol. 34, 2002, pp. 457-464.

[Ice76] H. W. Icenogle, B. C. Platt and W. L. Wolfe. Refractive indexes and temperature coefficients of germanium and silicon. *Applied Optics*, vol. 15, 1976, pp. 2348-2351.

[Jon89] R. Jones and C. Wykes. Holographic and Speckle Interferometry. Cambridge University Press, second edition, 1989.

[Jup88] W. P. O. Jüptner, T. Kreis and J. Geldmacher. Determination of absolute fringe order in hologram interferometry with wavelength controlled lasers. *SPIE Proc. Indust. Laser Interf. II*, vol. 955, 1988, pp.143-146.

[Kad97] H. Kadono, H. Takei and S. Toyooka. A noise-immune method of phase unwrapping in speckle interferometry. *Opt.&Lasers in Eng.*, vol. 26, 1997, pp. 151-164.

[Kel96] M. Kelly, G. Servais, T. Diep, D. Lin, S. Twerefour and G. Shah. A comparison of electrostatic discharge models and failure signatures for CMOS integrated circuit devices. *Journal of Electrostatics*, vol. 38, 1996, pp. 53.

[Kem03] Q. Kemao, S. H. Soon and A. Asundi. Smoothing filters in phase-shifting interferometry. *Opt.&Laser Tech.*, vol. 35, 2003, pp. 649-654.

[Kol92] J. Kölzer, C. Boit, A. Dallmann, G. Deboy, J. Otto and D. Weinmann. Quantitative emission microscopy. *J. Appl. Phys.*, vol. 71, no. 11, 1992, pp. R23-R41.

[Kol96] J. Kölzer, E. Oesterschulze and G. Deboy. Thermal imaging and measurement techniques for electronic materials and devices. *Microel. Eng.*, vol. 31, 1996, pp. 251-270.

[Kos88] N. Koskowich, R. B. Darling andf M. Soma. Effect of first-order phonon-assisted scattering on near-infrared free-carrier optical absorption in silicon. *Phys. Rev B.*, vol. 38, 1988, pp. 1281-1284.

[Kra92] J. Kraus, G. Sölkner and A. A. Valenzuela. An electro-optic laser probe at low temperature for the characterization of planar integrated microwave resonators. *Microelectr. Eng.*, vol. 16, 1992, pp. 333-340.

[Kre86] T. Kreis. Digital holographic interference-phase measurement using the Fourier-transform method. *J. Opt. Soc. Am. A*, vol. 3, 1986, pp. 847-855.

[Kre88] T. Kreis and W. Jüptner. Digital processing of holographic interference patterns using Fourier-transform methods. *Measurement*, vol.6, 1988, pp. 37-40.

[Kre91] T. Kreis and W. Osten. Automatische Rekonstruktion von Phaseverteilungen aus Interferogrammen. *Tech. Messen*, vol. 58, 1991, pp. 235-246.

[Kre96] T. Kreis. Holographic interferometry, Akademie V., Berlin, 1996.

[Kro93] R. Kropf, C. Russ, R. Kolbinger, H. Gieser and S. Irl. Zeitaufgelöste Untersuchung des Snapbackverhaltens eines ESD-Schutztransistors. *Tagungsband 3.ESD-Forum 1993*, 1993, pp. 19-26.

[Kru01] S. Krüger, G. Wernicke, W. Osten, D. Kayser, N. Demoli and H. Gruber. Fault detection and feature analysis in interferometric fringe patterns by the application of wavelet filters in convolution processors. *J. of Elect. Imaging*, vol. 10, 2001, pp. 228-233.

[Kru99] S. Krüger, L. Bouamama, H. Gruber, S. Teiwes and G. Wernicke. Analysis of interferometric fringe patterns by optical wavelet transform. *SPIE Proc. Optical Measur. Systems for Industrial Inspec.*, vol. 3824, 1999, pp. 222-228.

[Kru97] O. Kruschke, G. Wernicke, T. Huth, N. Demoli and H. Gruber. Holographic interferometric microscope for compete displacement determination. *Opt. Eng.*, vol. 36, 1997, pp. 2448-2457.

[Kud95] A. V. Kudryashov and A. V. Seliverstov. Adaptive stabilized interferometer with laser diode. *Opt. Commun.*, vol. 120, 1995, pp. 239-244.

[Kuj98] M. Kujawinska and W. Osten. Fringe pattern analysis methods: up to date review. *SPIE Proc.*, vol. 3407, 1998, pp. 56-65.

[Kwo87] O. Y. Kwon. Advanced wavefront sensing at lockheed. In N. A. Massie, ed., Interferometric Metrology, Proc. of Soc. Photo-Opt. Instr. Eng., vol. 816, 1987, pp. 196-211.

[Lim90] Lim, Jae S. Two-dimensional signal and image processing. Englewood Cliffs, NJ: Prentice Hall, 1990, pp. 536-540.

[Lit03] M. Litzenberger, C. Fürböck, S. Bychikhin, D. Pogany and E. Gornik. Scanning heterodyne interferometer setup for the time resolved thermal and free carrier mapping in semiconductor devices. *IEEE Transactions on Instrumentation and Measurement*, will be published in Oct or Dec 2005.

[Liu02] Z. Liu, M. Centurion, G. Panotopoulos, J. Hong and D. Psaltis. Holographic recording of fast events on a CCD camera. *Opt. Lett.*, vol. 27, 2002, pp. 22-24.

Bibliography

[Lun91a] M. Lundquist, H. Bleichner and E. Nordlander. An optical system for bilateral recombination-radiation diagnostics of the carrier redistribution in switching power devices. *IEEE Trans Instr. Meas.*, vol. 40, 1991, pp. 956-961.

[Lun91b] M. Lundquist, H. Bleichner and E. Nordlander. An optical system for bilateral recombination-radiation diagnostics of the carrier redistribution in switching power devices. *IEEE Trans Instr. Meas.*, vol. 40, 1991, pp. 956-961.

[Lut89] W. Luth. Isolation of moving objects in digital image sequences. Academy of Science of the GDR, 1989.

[Mal85] T. Maloney. Designing MOS inputs and outputs to avoid oxide failure in the charged device model. *Proc. EOS/ESD Symp.*, 1988, pp. 220-227.

[Mas79] N. A. Massie, R. D. Nelson and S. Holly. High-preformance real-time heterodyne interferometry. *Appl. Opt.*, vol. 18, 1979, pp. 1797-1803.

[McC94] J. A. McCaulley, V. M. Donnelly, M. Vernon, and I. Taha. Temperature dependence of the near-infrared refractive index of silicon, gallium arsenide, and indium phosphide. *Phys. Rev. B.*, vol. 49, 1994, pp. 7408-7417.

[Mei92] H. Meinke and F. W. Gundlach. Taschenbuch der Hochfrequenztechnik. Edited by K. Lange and K. H. Löcherer, Springer, 1992, pp. B13-B14.

[MIL-STD] MIL-STD 883.C method 3015.7 notice 8, Military standard for test methods and procedures for microelectronics: Electrostatic sensitivity classification. March 1989.

[Mus96] C. Musshoff, H. Wolf, H. Gieser, P. Egger and X. Guggenmos. Risetime effects of HBM and square pulses on the failure threshold of gg-nMOS transistors. *Microel. Reliab.*, vol. 36, 1996, pp. 1743-1746.

[Nie02] J. J. Niederhauser, D. Frauchinger, H.P. Weber and M. Frentz. Real time optoacoustic imaging using Schlieren transducer. *Appl. Phys. Lett.*, vol. 81, 2002, pp. 571.

[Otn78] R. K. Otnes and L. Enochson. Applied time series analysis. Wiley, New York, 1978.

[Pan98] M. Paniccia, R. M. Rao and W. M. Yee. Optical probing of flip chip packaged microprocessors. *J. Vac. Sci. Technol. B.*, vol. 16, 1998, pp. 3625-3630.

[Pog05] D. Pogany, S. Bychikhin, M. Denison, P. Rodin, N. Jensen, G. Groos, M. Stecher and E. Gornik. Thermally-driven motion of current filaments in ESD protection devices. *Solid-State Electronics*, vol. 49, 2005, pp. 421-429.

[Pog02b] D. Pogany, V. Dubec, S. Bychikhin, C. Fürböck, M. Litzenberger, G. Groos, M. Stecher and E. Gornik. Single-shot thermal energy mapping of semiconductor devices with the nanosecond resolution using holographic interferometry. *IEEE Electron Dev. Lett.*, vol. 23, 2002, pp. 606-608.

[Pog02c] D. Pogany, S. Bychikhin, C. Fürböck, M. Litzenberger, E. Gornik, G. Groos, K. Esmark and M. Stecher. Quantitative internal thermal energy mapping of semiconductor devices under short current stress using backside laser interferometry. *IEEE Trans. Electron Devices*, vol. 49, 2002, pp. 2070-2079.

[Pog03a] D. Pogany, V. Dubec, S. Bychikhin, C. Fürböck, M. Litzenberger, S. Naumov, G. Groos, M.Stecher and E. Gornik. Single-shot nanosecond thermal imaging of semiconductor devices using absorption measurements. *IEEE Trans. on Dev. and Mat. Reliab.*, vol. 3, no. 3, sep. 2003, pp. 85-88.

[Pog03b] D. Pogany, S. Bychikhin, J. Kuzmik, V. Dubec, N. Jensen, M. Denison, G. Groos, M. Stecher, E. Gornik. Thermal distribution during destructive pulses in ESD protection devices using a single-shot two-dimensional interferometric method. *IEEE Trans. on Dev. and Mat. Reliab.*, vol. 3, no. 4, 2003, pp. 197-201.

[Pog97] D. Pogany, C. Fürböck, N. Seliger, P. Habas, E. Gornik, S. Kubicek and S. Decoutere. Optical testing of submicron-technology MOSFETs and bipolar transistors. *Proc. ESSDERC'97*, Stuttgart, 1997, pp. 372-375.

[Pog98b] D. Pogany, N. Seliger, E. Gornik, M. Stoisiek and T. Lalinsky. Analysis of the temperature evolution from the time resolved thermooptical interferometric measurements with few Fabry-Perot peaks. *J. Appl. Phys.*, vol. 84, no. 8, 1998, pp. 4495-4501.

[Pog98a] D. Pogany, N. Seliger, T. Lalinsky, J. Kuzmik, P. Habas, P. Hrkut and E. Gornik. Study of thermal effects in Ga as micromachined power sensor microsystems by an optical interferometer technique. *Microel. J.*, vol. 29, 1998, pp. 191-198.

[Pog02a] D. Pogany, S. Bychikhin, M. Litzenberger, E. Gornik, G. Groos, M. Stecher. Extraction of spatio-temporal distribution of power dissipation in semiconductor devices using nanosecond interferometric mapping technique. *Appl. Phys. Lett.*, vol. 81, 2002, pp. 2881-2883.

[Pog03c] D. Pogany, S. Bychikhin, E. Gornik, M. Denison, N. Jensen, G. Groos and M. Stecher. Moving current filaments in ESD protection devices and their relation to electrical characteristics. *Proc. IEEE Int. Reliab. Phys. Symp (IRPS 2003)*, 2003, pp. 241-248.

Bibliography

[Pre92] W. H. Press, S. A. Teukolsky, W. T. Vetterling and B. P. Flannery. Numerical recipes in Fortran. Cambridge Univ. press, 1992.

[Qua02] C. Quan, C. J. Tay, X. Y. He, X. Kang and H. M. Shang. Microscopic surface contouring by fringe projection metod. *Opt.&Laser Techn.*, vol. 34, 2002, pp. 547-552.

[Ras94] P. K. Rastogi. Techniques to measure displacement, derivatives and surface shapes. Extension to comparative holography. In P. K. Rastogi, ed., Holographic Interferometry, Springer Series in Optical Sciences, 68, 1994, pp. 213-292.

[Rei95] J. C. Reiner. Latent gate oxide damage by ultra-fast electrostatic discharge. Hartung-Gorre Verlag, Konstanz, Germany, 1995.

[Rem03] M. Remmach, R. Desplats, F. Beaudoin, E. Frances, P. Pedru and D. Lewis. Time resolved photoemission (PICA) – from the physics to practical considerations. *Microel. Reliab.*, vol. 43, 2003, pp. 1639-1644.

[Ric00] S. Richter, M. Geva, J. P. Garno and R. N. Kleiman. Metal insulator semiconductor tunneling microscope, two dimmensional dopant profiling of semiconductors with conducting atomic force microscopy. *Appl. Phys. Lett.* vol. 77, 2000, pp. 456.

[Rob93] D. W. Robinson and G. T. Reid. Interferogram analysis. Inst. of Phys. Publ., Bristol, 1993.

[Rog96] H. Rogne, P. J. Timans and H. Ahmed. Infrared absorption in silicon at elevated temperatures. *Appl. Phys. Lett.*, vol. 69, 1996, pp. 2190-2192.

[Rus98] C. Russ, K. Bock, M. Rasras, I. De Wolf, G. Groeseneken and H. E. Maes. Non-uniform triggering of gg-NMOSt investigated by combined emission microscopy and transmission line pulsing. *Proc. EOS/ESD Symp.*, 1998, pp.177-186.

[Sal91] B. E. A. Saleh and M. C. Teich. Fundamentals of Photonics. Wiley, 1991.

[Sav64] A. Savitzky and M. J. E. Golay. Smoothing and differential of data by simplified least square procedure. *Analytical Chemistry*, vol. 36, 1964, pp.1627-1639.

[Sch01] Schlieren method for imaging of semiconductor device parameters. US patent US6181416, 2001

[Sch95] U. Schnars, T. Kreis and W. P. O. Jüptner. CCD recording and numerical reconstruction of holograms and holographic interferograms. *SPIE Proc. Interferometry VII: Techniques and Analysis*, vol. 2544, 1995, pp.57-63.

Bibliography

[Sel97] N. Seliger, P. Habas, D. Pogany and E. Gornik. Time-resolved analysis of self-heating in power VDMOSFETs using backside laser probing. *Solid-State Electronics*, vol. 41, 1997, pp. 1285-1292.

[Sod95] J. M. Soden and R. E. Anderson. IC failure analysis: Techniques and tools for quality and reliability improvement. *Microel. Reliab.*, vol. 33, 1995, pp.429-453.

[Sor87] R. A. Soref and B. R. Bennett. Electrooptical effects in silicon. *IEEE J. Quant. Electron.*, vol. 23, 1987, pp. 123-129.

[Stu92] J. C. Sturm and C. M. Reaves. Silicon temperature measurement by infrared absorption: fundamental processes and doping effects. *IEEE Trans. Electron Devices*, vol. 39, no. 1, 1992, pp. 81-88.

[Tak82] M. Takeda, H. Ina and S. Kobayashi. Fourier-transform method of fringe-pattern analysis for computer-based topography and interferometry. *J. Opt. Soc. Am.*, vol.72, 1982, pp.156-160.

[Tha98] R. Thalhammer, C. Fürböck, N. Seliger. G. Deboy, E. Gornik and G. Wachutka. Internal characterization of IGBT using the backside laser probing technique - Interpretation of measurements by numerical simulation. *Proc. ISPSD'98*, Kyoto, Japan, 1998, pp. 199 - 202.

[Tim93] P. J. Timans. Emissivity of silicon at elevated temperatures. *J. Appl. Phys.*, vol. 74, 1993, pp. 6353-6364.

[Var98] J. Varesi and A. Majumdar. Scanning Joule expansion microscopy at nanometer scales. *Appl. Phys. Lett.*, vol. 72, 1998, pp. 37-39.

[Ver94] P. Verguin. Liquid crystals in failure analysis today. *Microelectr. Eng.*, vol. 24, 1994, pp. 211-218.

[Vro91] H. A. Vrooman. Quaint: quantitative analyses of interferograms. Delft Univ. Press, 1991.

[Wol96] I. Wolf. Micro-Raman spectroscopy to study local mechanical stress in silicon integrated circuits. *Semicond. Sci. Technol.*, vol. 11, 1996, pp. 139-154.

[Yam96] I. Yamaguchi, J.-Y. Liu, T. Nakajima and J. Kato. Active stabilization and real time analysis of interference fringes. *SPIE Proc. Optical Inspection and Micromeasurements*, vol. 2782, 1996, pp. 354-362.

List of publications and conference contributions

1. V. Dubec and P. Vojtek. Impulzný režim generácie CO_2 lasera. (The pulsed mode operation of a CO_2 laser), in slovak language, *Jemná mechanika a optika*, vol. 7-8, 2001, pp. 232-234.
2. V. Dubec, P. Vojtek. Pulse lenght measurement of externally modulated CO_2 laser. *Proc. Measurement 2001*, Smolenice, Slovakia, May 14-17, 2001, pp. 410-413.
3. D. Pogany, V. Dubec, S. Bychikhin, C. Fürböck, M. Litzenberger, G. Groos, M. Stecher and E. Gornik. Single-shot thermal energy mapping of semiconductor devices with the nanosecond resolution using holographic interferometry. *IEEE Electron Dev. Lett.*, vol. 23, 2002, pp. 606-608.
4. S. Bychikhin, V. Dubec, M. Litzenberger, D. Pogany, E. Gornik, G. Groos, K. Esmark, W. Stadler, H. Gieser and H. Wolf. Investigation of ESD protection elements under high current stress in CDM-like time domain using backside laser interferometry. *Proc. 24th Electrical Overstress/ Electrostatic Discharge Symposium (EOS/ESD 2002)*, Charlotte, USA, 2002, pp. 387-395.
5. D. Pogany, S. Bychikhin, J. Kuzmik, V. Dubec, N. Jensen, M. Denison, G. Groos, M. Stecher and E. Gornik. Investigation of thermal distribution during destructive pulses in ESD protection devices using a single-shot, two-dimensional interferometric method. *IEDM 2002 Technical Digest (IEEE International Electron Device Meeting)*, San Francisco, USA, 2002, pp. 345-348.
6. D. Pogany, S. Bychikhin, M. Blaho, V. Dubec, J. Kuzmik, M. Litzenberger, C. Pflügl, G. Strasse and E. Gornik. Transient interferometric mapping of temperature and free carriers in semiconductor devices. *Proc. IEEE-LEOS 2003*, 2003, pp. 666-667.
7. M. Blaho, M. Denison, V. Dubec, D. Pogany, M. Stecher and E. Gornik. Hot spot mapping in the DMOS devices for automotive applications. *Beiträge der Informationstagung Mikroelektronik*, 2003, pp. 329-334.
8. M. Denison, M. Blaho, D. Silber, J. Joos, M. Stecher, V. Dubec, D. Pogany and E. Gornik. Hot Spot Dynamics in Quasi Vertical DMOS under ESD Stress. *The 15th International*

List of publications and conference contributions

Symposium on Power Semiconductor Devices & ICs (*ISPSD'03*), Cambridge, England, pp.80-83.

9. S. Bychikhin, V. Dubec, M. Litzenberger, D. Pogany, E. Gornik, G. Groos, K. Esmark, M. Stecher, W. Stadler, H. Gieser and H. Wolf. Investigation of ESD protection elements under high current stress in CDM-like time domain using backside laser interferometry. *J. of Electrostatic*, vol. 59, 2003, pp. 241-255.

10. M. Denison, M. Blaho, D. Silber, J. Joost, N. Jensen, M. Stecher, V. Dubec, D. Pogany and E. Gornik. Hot spot dynamics in quasi vertical DMOS under ESD stress. *Proc. ISPSD'03*, 2003, pp. 80-83.

11. V. Dubec, S. Bychikhin, M. Blaho, D. Pogany, E. Gornik, J. Willemen, N. Qu, W. Wilkening, L. Zullino and A. Andreini. A dual-beam Michelson interferometer for investigation of trigger dynamics in ESD protection devices under very fast TLP stress. (*ESREF'03*) *Microel. Reliab.*, vol. 43, 2003, pp. 1557-1561.

12. D. Pogany, V. Dubec, S. Bychikhin, C. Fürböck, M. Litzenberger, S. Naumov, G. Groos, M.Stecher and E. Gornik. Single-shot nanosecond thermal imaging of semiconductor devices using absorption measurements. *IEEE Trans. on Dev. and Mat. Reliab.*, vol. 3, no. 3, 2003, pp. 85-88.

13. D. Pogany, S. Bychikhin, J. Kuzmik, V. Dubec, N. Jensen, M. Denison, G. Groos, M. Stecher, E. Gornik. Thermal distribution during destructive pulses in ESD protection devices using a single-shot two-dimensional interferometric method. *IEEE Trans. on Dev. and Mat. Reliab.*, vol. 3, no. 4, 2003, pp. 197-201.

14. M. Denison, M. Blaho, P. Rodin, V. Dubec, D. Pogany, D. Silber, E. Gornik and M. Stecher. Moving current filaments in integrated DMOS transistors under short-duration current stress. *IEEE Trans. Electron Dev.*, vol. 51, 2004, pp. 1695-1703.

15. V. Dubec, S. Bychikhin, M. Blaho, M. Heer, D. Pogany, M. Denison, N. Jensen, M. Stecher, G. Groos and E. Gornik. Multiple-time-instant 2D thermal mapping during a single ESD event. (*ESREF'04*) *Microel. Reliab.*, vol. 44, 2004, pp. 1793-1798.

16. V. Dubec, S. Bychikhin, D. Pogany, E. Gornik, G. Groos and M. Stecher. Error analysis in phase extraction in a 2D holographic imaging of semiconductor devices. *Proc. SPIE, Practical Holography XVIII: Materials and Applications*, vol. 5290, 2004, pp. 233-242.

List of publications and conference contributions

17. S. Bychikhin, V. Dubec, D. Pogany, E. Gornik, M. Graf, V. Dudek and W Soppa. Transient interferometric mapping of smart power SOI ESD protection devices under TLP and vf-TLP stress. (*ESREF'04*) *Microel. Reliab.*, vol. 44, 2004, pp. 1687-1692.
18. M. Denison, M. Blaho, P. Rodin, V. Dubec, D. Pogany, D. Silber, E. Gornik and M. Stecher. Moving current filaments in integrated DMOS transistors under short-duration current stress. *IEEE Trans. on Electron Devices*, vol. 51, 2004, pp. 1331-1339.
19. V. Dubec, S. Bychikhin, M. Blaho, M. Heer, D.Pogany, E. Gornik, M. Denison, M. Stecher and G.Groos. Thermal imaging at multiple time instants for study of self-heating and ESD phenomena. *Proceeding GMe Forum*, Vienna, Austria, 2005, ISBN 3-901578-15-3, pp. 167-171.
20. J. Kuzmík, S Bychikhin, V. Dubec, M. Blaho, M. Marso, P. Kordoš, T. Suski, M. Bockowski, I Grzegory and D. Pogany. Characterization of III-Nitride Group Semiconductors and Devices Using Optical Methods. *The 29th Workshop on Compound Semicond. Devices and Integrated Circuits held in Europe WOCSDICE 05*, Cardiff, 2005, pp. 61-62.
21. M. Heer, V. Dubec, M. Blaho, S. Bychikhin, D. Pogany and E. Gornik. Automated setup for thermal imaging and electrical degradation study of power DMOS devices," accepted for *ESREF'05*.

List of acronyms

2D – two-dimensional
AC – alternating current
BLI – backside laser interferometry
CCD – charge-coupled device
CDM – charged device model
DC – direct current
DET – optical detector
DFB – distributed feedback
DMOS – double diffused metal-oxide-semiconductor
DPSS – diode pumped solid state
DUT – device under test
ESD – electrostatic discharge
FFT – Fast Fourier Transform
FPA – focal plane array
FWHM – full width at half of maximum
HBM – human body model
HV – high voltage
IMAQ – image acquisition
IR – infra-red
IV – current-voltage characteristic
LED – light emitting diode
MI – Michelson interferometer
MO – microscope objective
NPBS – non-polarising beam splitter
OBIC – optical beam induced current
OPO – optical parametric oscillator
PAL – phase alternating line
PBS – polarising beam splitter

List of acronyms

S/N – signal-noise ratio
SPT – smart power technology
TIM – transient interferometric mapping
TL – transmission line
TLP – transmission line pulse
TTL – transistor-transistor logic
vf-TLP – very-fast TLP

List of frequently used symbols

Symbol	Unit	Description
$A(x, y)$	[a.u.]	background intensity
$a(u, v)$	[a.u.]	background intensity spectrum
α	[m^{-1}]	absorption coefficient
$B(x, y)$	[a.u.]	fringe amplitude (real quantity)
β	-	tolerance factor
$C(x, y)$	[a.u.]	fringe amplitude (complex quantity)
$c(u, v)$	[a.u.]	spectrum of C
d_{difr}	[m]	diffraction limit
E	[Vm^{-1}]	amplitude of the laser beam
f_F	[a.u.]	fringe (spatial) carrier frequency
f_c	[a.u.]	Nyquist frequency
φ	[rad]	phase of the laser beam
$\Delta\varphi$	[rad]	stress-induced optical phase shift
ϕ	[rad]	optical phase distribution
γ	-	degree of coherence
Γ	-	gamma characteristic
$i(u, v)$	[a.u.]	spectrum of the interferogram
I_0	[Wm^{-2}]	incident light intensity
I	[Wm^{-2}]	light intensity, interferogram intensity distribution
I_p	[Wm^{-2}]	probe beam intensity
I_r	[Wm^{-2}]	reference beam intensity
I_{DUT}	[A]	current through the DUT
$\vec{k}$	[m^{-1}]	wave vector
κ	[WK^{-1}m^{-1}]	thermal conductivity
L	[m]	substrate thickness
L_{coh}	[m]	laser coherence length

List of frequently used symbols

L_{th}	[m]	thermal diffusion length
L_{dev}	[m]	DUT size (length, width)
λ	[m]	light wavelength
n	-	refractive index
Δn	-	refractive index change
N_A	-	numerical aperture
N	-	number of pixels in interferogram
P_{2D}	[Wm^{-2}]	instantaneous 2D power dissipation density
Σ	[WK^{-1}m^{-3}]	time derivative of phase shift
T_0	[K]	ambient temperature
τ_{TLP}	[s]	TLP pulse length
ω	[s^{-1}]	light frequency
Ψ	[WK^{-1}m^{-3}]	space derivatives of phase shift

Acknowledgement

This work was supported by European Community project DEMAND IST-2000-30033, by the Austrian Science Fund under the FWF-Wittgenstein award, by the Austrian Microelectronics Society (GMe), by the European Community Medea+ Project T102 "ASDESE" and by the Austrian government.

Südwestdeutscher Verlag
für Hochschulschriften

Wissenschaftlicher Buchverlag bietet
kostenfreie
Publikation
von
Dissertationen und Habilitationen

Sie verfügen über eine wissenschaftliche Abschlußarbeit zu aktuellen oder zeitlosen Fragestellungen, die hohen inhaltlichen und formalen Ansprüchen genügt, und haben **Interesse an einer honorarvergüteten Publikation?**

Dann senden Sie bitte erste Informationen über Ihre Arbeit per Email an: info@svh-verlag.de.

Unser Außenlektorat meldet sich umgehend bei Ihnen.

Südwestdeutscher Verlag für Hochschulschriften
Aktiengesellschaft & Co. KG
Dudweiler Landstr. 99
D – 66123 Saarbrücken
www.svh-verlag.de

Printed by Books on Demand GmbH, Norderstedt / Germany